〔大数据技术丛书〕

Hadoop+Spark
大数据分析实战

迟殿委 编著

清华大学出版社
北京

内 容 简 介

本书是 Hadoop + Spark 大数据分析技术入门书，基于 Hadoop 和 Spark 两大框架体系的 3.2 版本，以通俗易懂的方式介绍 Hadoop + Spark 原生态组件的原理、集群搭建、实战操作，以及整个 Hadoop 生态系统主流的大数据分析技术。

本书共分 14 章。第 1 章讲解 Hadoop 框架及新版本特性，并详细讲解大数据分析环境的搭建工作，包括 Linux 操作系统的安装、SSH 工具使用和配置等；第 2 章讲解 Hadoop 伪分布式的安装和开发体验，使读者熟悉 Hadoop 大数据开发两大核心组件，即 HDFS 和 MapReduce；第 3~12 章讲解 Hadoop 生态系统各框架 HDFS、MapReduce、输入输出、Hadoop 集群配置、高可用集群、HBase、Hive、数据实时处理系统 Flume，以及 Spark 框架数据处理、机器学习等实战技术，并通过实际案例加深对各个框架的理解与应用；第 13~14 章分别通过影评分析、旅游酒店评价分析实战项目来贯穿大数据分析的完整流程。

本书可以作为大数据分析初学者的入门指导书，也可以作为大数据开发人员的参考手册，同时也适合作为高等院校大数据相关专业的教材或教学参考书。

本书封面贴有清华大学出版社防伪标签，无标签者不得销售。
版权所有，侵权必究。举报：010-62782989，beiqinquan@tup.tsinghua.edu.cn。

图书在版编目（CIP）数据

Hadoop+Spark 大数据分析实战 / 迟殿委编著. —北京：清华大学出版社，2022.6（2025.1 重印）
（大数据技术丛书）
ISBN 978-7-302-60884-4

Ⅰ. ①H… Ⅱ. ①迟… Ⅲ. ①数据处理软件 Ⅳ. ①TP274

中国版本图书馆 CIP 数据核字（2022）第 083224 号

责任编辑：夏毓彦
封面设计：王　翔
责任校对：闫秀华
责任印制：宋　林

出版发行：清华大学出版社
网　　址：https://www.tup.com.cn，https://www.wqxuetang.com
地　　址：北京清华大学学研大厦 A 座　　邮　编：100084
社 总 机：010-83470000　　邮　购：010-62786544
投稿与读者服务：010-62776969，c-service@tup.tsinghua.edu.cn
质 量 反 馈：010-62772015，zhiliang@tup.tsinghua.edu.cn

印 装 者：小森印刷霸州有限公司
经　　销：全国新华书店
开　　本：190mm×260mm　　印　张：18.75　　字　数：506 千字
版　　次：2022 年 7 月第 1 版　　印　次：2025 年 1 月第 3 次印刷
定　　价：69.00 元

产品编号：096183-01

前　言

如今大数据技术已广泛应用于金融、医疗、教育、电信、政府等领域。各个行业都积累了大量的历史数据，并不断产生大量新数据，数据的种类不断增多，数据体量也急剧增长，数据计量单位已经发展到 PB、EB、ZB、YB 级甚至 BB、NB、DB 级，传统的数据存储、管理、分析技术已经无法满足大数据的处理要求。大数据分析不同于传统的数据处理方式，需要通过分布式存储和分布式运算来实现，这也催生了优秀的大数据处理框架和生态组件的出现，Hadoop 便是最具代表性的大数据处理生态系统框架，Spark 则是更为高效的数据处理框架，二者的结合可以为大数据分析和机器学习提供可靠且高效的解决方案。许多大型互联网公司，如谷歌、阿里巴巴、百度、京东等都急需掌握大数据技术人才，大数据技术人才出现了供不应求的状况。

写作思路

本书从大数据开发和大数据分析岗位需求出发，力求从 Hadoop 生态圈和 Spark 生态系统全面解析每个组件。Hadoop 框架方面，包括大数据平台搭建、Hadoop 各典型组件的实战应用、新版本的集群配置和高可用特性、Hive 和 HBase 的搭建与实战等。Spark 框架方面，包括 Spark 框架数据处理等的基础知识、机器学习实战应用、集群环境搭建，同时包括常用的 Shell 命令、API 操作。本书最后安排了两个综合项目实战案例，一方面用来对 Hadoop+Spark 框架进行大数据开发和大数据分析的基础内容进行巩固和提高，另一方面，结合电影评论分析和旅游评论分析这样的实际场景，使读者能够把握真实的大数据开发或大数据分析应用项目的技术内容，从而对大数据分析的典型流程有清晰的理解，完成从数据采集、数据分析到数据可视化各个环节的全面掌握。全书实战操作和应用案例丰富，每一个知识点都讲得十分细致，让读者能够轻松地步入大数据开发工程师的大门。

关于本书

本书是一本关于大数据平台应用和大数据分析方面的实战书籍，知识面比较广，涵盖整个 Hadoop 生态系统主流的大数据开发技术，以及用于数据实时处理的 Spark 框架。力从实践操作讲起，尽量去除那些影响读者理解的纯理论内容。等基本的操作已经掌握以后，再回过头来讲解与实战相关的核心理论知识。所以，本书讲解的方法是先动手实践再理解理论。首先基于目前新版本的 Hadoop 框架展开，采用 Hadoop 3.2.2 版本阐述大数据平台搭建和开发方面的内容。随着 Hadoop 生态系统的成长，Hadoop 已经不再是一个简单的数据分布式存储平台和工具，已经形成一个完整的Hadoop生态圈。本书全面讲解Hadoop生态圈各组件的核心知识和操作方法。然后，系统介绍Spark框架搭建和操作，并结合经典的机器学习算法，讲解基于Spark平台的大数据分析技术。最后通过两个综合实战项目来体现大数据分析的完整流程。

本书采用先上手实践，后归纳和学习理论知识的思路编写，读者可以快速上手基于Hadoop和Spark的大数据开发应用，读者还可以对照书中的步骤成功搭建属于自己的大数据集群，并独立完成项目开发。书中提供了大数据分析的详细步骤，并配套了源代码。

本书内容

全书共14章，第1章讲解Hadoop框架及新版本特性，并详细讲解大数据环境的准备工作，包括Linux操作系统的安装、SSH工具使用和配置等；第2章讲解Hadoop伪分布式的安装和开发体验，使读者熟悉Hadoop大数据开发两大核心组件，即HDFS和MapReduce；第3~12章讲解Hadoop生态系统各框架HDFS、MapReduce、输入输出、Hadoop集群配置、HA高可用集群、HBase、Hive、数据实时处理系统Flume、Spark框架数据处理等的基础知识、机器学习实战应用、集群环境搭建，同时包括常用的Shell命令、API操作等，并通过实际操作加深对各个框架的理解与应用；第13~14章分别通过影评分析、旅游酒店评价分析实战项目巩固所学知识，案例涉及自然语言处理和数据可视化入门内容，使读者掌握的技术更加全面。

配套资源下载

本书配套资源包括源码、PPT课件、开发环境、答疑服务，可用微信扫描下面的二维码获取，也可按扫描后的页面提示把下载链接转发到自己的邮箱中下载。如果有疑问和建议，请联系booksaga@163.com，邮件主题为"Hadoop+Spark大数据分析实战"。

适合阅读本书的读者

本书可作为大数据分析初学者的入门指导书、大数据开发人员的参考用书，也可以作为高校大数据平台搭建或大数据开发课程的参考教材。学习本书要求读者有一定的Java编程基础，并掌握Linux系统的基础知识。

作　者
2022年4月

目 录

第1章 大数据与 Hadoop ··1
- 1.1 什么是大数据 ··1
- 1.2 大数据的来源 ··2
- 1.3 如何处理大数据 ··3
  - 1.3.1 数据分析与挖掘 ··3
  - 1.3.2 基于云平台的分布式处理 ··4
- 1.4 Hadoop 3 新特性 ···6
- 1.5 虚拟机与 Linux 操作系统的安装 ···7
  - 1.5.1 VirtualBox 虚拟机的安装 ··7
  - 1.5.2 Linux 操作系统的安装 ···8
- 1.6 SSH 工具与使用 ···14
- 1.7 Linux 统一设置 ··16
- 1.8 本章小结 ··17

第2章 Hadoop 伪分布式集群 ···18
- 2.1 安装独立运行的 Hadoop ··19
- 2.2 Hadoop 伪分布式环境准备 ···21
- 2.3 Hadoop 伪分布式安装 ···25
- 2.4 HDFS 操作命令 ··31
- 2.5 Java 项目访问 HDFS ···33
- 2.6 winutils ··38
- 2.7 快速 MapReduce 程序示例 ··39
- 2.8 本章小结 ··42

第3章 HDFS 分布式文件系统 ··43
- 3.1 HDFS 的体系结构 ···43
- 3.2 NameNode 的工作 ··44
- 3.3 SecondaryNameNode ···49
- 3.4 DataNode ··50

3.5 HDFS 的命令 ·· 51
3.6 RPC 远程过程调用 ·· 52
3.7 本章小结 ··· 53

第 4 章 分布式运算框架 MapReduce ·· 55
4.1 MapReduce 的运算过程 ··· 55
4.2 WordCount 示例 ··· 57
4.3 自定义 Writable ··· 60
4.4 Partitioner 分区编程 ·· 63
4.5 自定义排序 ·· 65
4.6 Combiner 编程 ··· 67
4.7 默认 Mapper 和默认 Reducer ·· 68
4.8 倒排索引 ··· 69
4.9 Shuffle ··· 73
4.9.1 Spill 过程 ·· 73
4.9.2 Sort 过程 ·· 74
4.9.3 Merge 过程 ·· 75
4.10 本章小结 ·· 76

第 5 章 Hadoop 输入输出 ·· 78
5.1 自定义文件输入流 ·· 79
5.1.1 自定义 LineTextInputFormat ··· 79
5.1.2 自定义 ExcelInputFormat 类 ··· 82
5.1.3 DBInputFormat ··· 86
5.1.4 自定义输出流 ·· 89
5.2 顺序文件 SequenceFile 的读写 ··· 90
5.2.1 生成一个顺序文件 ·· 91
5.2.2 读取顺序文件 ·· 91
5.2.3 获取 Key/Value 类型 ··· 92
5.2.4 使用 SequenceFileInputFormat 读取数据 ························· 93
5.3 本章小结 ·· 95

第 6 章 Hadoop 分布式集群配置 ·· 96
6.1 Hadoop 集群 ·· 96

6.2	本章小结	100

第 7 章 Hadoop 高可用集群搭建 ... 101

7.1	ZooKeeper 简介	101
7.2	ZooKeeper 集群安装	104
7.3	znode 节点类型	105
7.4	观察节点	106
7.5	配置 Hadoop 高可靠集群	106
7.6	用 Java 代码操作集群	115
7.7	本章小结	117

第 8 章 数据仓库 Hive ... 118

8.1	Hive 简介	118
8.2	Hive3 的安装配置	120
	8.2.1 使用 Derby 数据库保存元数据	120
	8.2.2 使用 MySQL 数据库保存元数据	121
8.3	Hive 命令	124
8.4	Hive 内部表	127
8.5	Hive 外部表	128
8.6	Hive 表分区	128
	8.6.1 分区的技术细节	128
	8.6.2 分区示例	131
8.7	查询示例汇总	133
8.8	Hive 函数	134
	8.8.1 关系运算符号	135
	8.8.2 更多函数	136
	8.8.3 使用 Hive 函数实现 WordCount	138
8.9	本章小结	140

第 9 章 HBase 数据库 ... 141

9.1	HBase 的特点	141
	9.1.1 HBase 的高并发和实时处理数据	142
	9.1.2 HBase 的数据模型	142
9.2	HBase 的安装	144

	9.2.1 HBase 的单节点安装	145
	9.2.2 HBase 的伪分布式安装	147
	9.2.3 Java 客户端代码	149
	9.2.4 其他 Java 操作代码	152
9.3	HBase 集群安装	155
9.4	HBase Shell 操作	159
	9.4.1 DDL 操作	160
	9.4.2 DML 操作	162
9.5	本章小结	166

第 10 章 Flume 数据采集ㆍㆍㆍ167

10.1	Flume 简介	167
	10.1.1 Flume 原理	167
	10.1.2 Flume 的一些核心概念	168
10.2	Flume 的安装与配置	169
10.3	快速示例	169
10.4	在 ZooKeeper 中保存 Flume 的配置文件	171
10.5	Flume 的更多 Source	174
	10.5.1 avro source	174
	10.5.2 thrift source 和 thrift sink	178
	10.5.3 exec source	181
	10.5.4 spool source	182
	10.5.5 HDFS sinks	183
10.6	本章小结	184

第 11 章 Spark 框架搭建及应用ㆍㆍㆍ185

11.1	安装 Spark	186
	11.1.1 本地模式	186
	11.1.2 伪分布式安装	188
	11.1.3 集群安装	191
	11.1.4 Spark on YARN	193
11.2	使用 Scala 开发 Spark 应用	196
	11.2.1 安装 Scala	196
	11.2.2 开发 Spark 程序	197

- 11.3 spark-submit ·· 200
  - 11.3.1 使用 spark-submit 提交 ·································· 200
  - 11.3.2 spark-submit 参数说明 ···································· 201
- 11.4 DataFrame ··· 203
  - 11.4.1 DataFrame 概述 ·· 203
  - 11.4.2 DataFrame 基础应用 ······································ 205
- 11.5 Spark SQL ·· 210
  - 11.5.1 快速示例 ··· 211
  - 11.5.2 Read 和 Write ·· 215
- 11.6 Spark Streaming ·· 216
  - 11.6.1 快速示例 ··· 217
  - 11.6.2 DStream ·· 220
  - 11.6.3 FileStream ··· 220
  - 11.6.4 窗口函数 ··· 222
  - 11.6.5 updateStateByKey ··· 223
- 11.7 共享变量 ··· 225
  - 11.7.1 广播变量 ··· 225
  - 11.7.2 累加器 ··· 227
- 11.8 本章小结 ··· 227

第 12 章 Spark 机器学习 ·· 228

- 12.1 机器学习 ··· 228
  - 12.1.1 机器学习概述 ··· 228
  - 12.1.2 Spark ML ··· 230
- 12.2 典型机器学习流程介绍 ··· 230
  - 12.2.1 提出问题 ··· 230
  - 12.2.2 假设函数 ··· 231
  - 12.2.3 代价函数 ··· 232
  - 12.2.4 训练模型确定参数 ··· 233
- 12.3 经典算法模型实战 ··· 233
  - 12.3.1 聚类算法实战 ··· 233
  - 12.3.2 回归算法实战 ··· 236
  - 12.3.3 协同过滤算法实战 ··· 239

第13章 影评分析项目实战 245

13.1 项目内容 245
13.2 项目需求及分析 246
13.3 详细实现 250
13.3.1 搭建项目环境 250
13.3.2 编写爬虫类 253
13.3.3 编写分词类 255
13.3.4 第一个 job 的 Map 阶段实现 259
13.3.5 一个 job 的 Reduce 阶段实现 259
13.3.6 第二个 job 的 Map 阶段实现 260
13.3.7 第二个 job 的自定义排序类阶段的实现 261
13.3.8 第二个 job 的自定义分区阶段实现 261
13.3.9 第二个 job 的 Reduce 阶段实现 262
13.3.10 Run 程序主类实现 262
13.3.11 编写词云类 263
13.3.12 效果测试 264

第14章 旅游酒店评价分析项目实战 266

14.1 项目介绍 266
14.2 项目需求及分析 267
14.2.1 数据集需求 267
14.2.2 功能需求 267
14.3 详细实现 268
14.3.1 数据集上传到 HDFS 269
14.3.2 Spark 数据清洗 271
14.3.3 构建 Hive 数据仓库表 274
14.3.4 Hive 表数据导出到 MySQL 280
14.3.5 数据可视化开发 282

第 1 章

大数据与 Hadoop

本章主要内容：

- 大数据概念及来源。
- 大数据处理方式介绍。
- Hadoop 简介。
- 虚拟机的安装与配置。
- Linux 的操作系统的安装。
- SSH（Secure Shell）。

本章首先介绍大数据的基础知识，包括大数据的相关概念和典型处理方式，然后详细介绍 Hadoop 框架。Hadoop 是一个由 Apache 基金会开发的分布式系统基础架构。Hadoop 的作者为 Doug Cutting，照片如图 1-1 所示，他也是 Lucene、Nutch 等项目的创始人。2004 年，Cutting 基于 Google（谷歌）发布的关于 GFS（Google File System）的学术文献打造出了 Hadoop。"Hadoop"并不是一串英文单词的首字母缩写，更没有任何的意义，这只是 Cutting 的孩子给自己的黄色毛绒小象玩具起的名字。Hadoop 可以读作：[hædu:p]。

图 1-1

Hadoop 的特点在于，用户可以在不了解分布式底层细节的情况下编写分布式程序，充分利用集群的威力进行高速运算和存储。

Hadoop 实现了一个分布式文件系统（Hadoop Distributed File System，简称 HDFS）。HDFS 有高容错性的特点，并且设计用来部署在低廉（low-cost）的硬件上，而且它提供高吞吐量（high throughput）来访问应用程序的数据，适合那些有着超大数据集（large data set）的应用程序。HDFS 放宽了可移植操作系统接口的要求，可以以流的形式访问文件系统中的数据。

Hadoop 框架最核心的设计就是 HDFS 和 MapReduce。HDFS 为海量的数据提供了存储，则 MapReduce 为海量的数据提供了计算。

1.1 什么是大数据

想了解什么是大数据？首先需要知道什么是数据，其实文本、声音、图片、视频都是数据。例如你用手机数据线连上电脑的时候上传的都是数据。

那么大数据如何定义呢？大数据指的就是数据体量达到了一定的级别，而我们现有的算法和工具无法在合理的时间内给予处理，这样的数据才可以称为大数据。当然，大数据还包括多样性（Variety）、价值密度低（Valueless）、处理速度快（Velocity）等特点。但最重要的特点还是数据量（Volume）要大。我们知道描述一个物品很大的时候是需要带上单位的。比方说，姚明很高，身高 230。这样描述显然不准确，是 cm（厘米）还是 mm（毫米），如果 230mm，那它只是个模型。同样道理，大数据也需要带有度量单位，下面是一些数据单位之间的换算关系。

- 1B（Byte，字节）=8bit
- 1KB（Kilobyte，千字节）=1024B
- 1MB（Mega byte，兆字节，简称兆）=1024KB
- 1GB（Giga byte，吉字节，又称千兆）=1024MB
- 1TB（Tera byte，万亿字节，太字节）=1024GB，其中 1024=2^{10}（2 的 10 次方）
- 1PB（Peta byte，千万亿字节，拍字节）=1024TB
- 1EB（Exa byte，百亿亿字节，艾字节）=1024PB
- 1ZB（Zetta byte，十万亿亿字节，泽字节）= 1024 EB
- 1YB（Yotta byte，一亿亿亿字节，尧字节）= 1024 ZB
- 1BB（Bronto byte，一千亿亿亿字节）= 1024 YB
- 1NB（Nona byte）= 1024BB
- 1DB（Dogga byte）= 1024NB

大家使用迅雷下载电影，下载速度显示的 500KB，B 指的就是基本单位，即字节 byte。其实大家对 KB、MB、GB 应该都是有一定的概念，例如使用手机拍一幅帅照大约 1MB 左右，一部电影差不多是几个 GB，甚至大家对 TB 也有概念，大家现在买移动硬盘基本都是 TB 级的容量了。而真正的大数据是需要至少达到这些单位的级别的，比如 PB、EB、ZB、YB、NB 等。

其实，这些单位是为大数据而生的，本来没有这些单位。1PB 就相当于美国国家图书馆藏书的所有内容之和。而 Google 每天都在处理 20PB 的数据。一般认为达到 PB 级别的数据才可以称为大数据。这里最大的单位是 YB，有家统计机构给出 1YB 相当于世界上所有海滩上的沙子粒数总和，准不准确无法验证，这只是说明数据体量达到了一个海量的级别。当然，还有更大的单位没有列出，比如比 YB 更大还有 NB，等等，数据增长不停止的话，单位定义不会停止。

1.2 大数据的来源

大数据一共有三个来源。第一个来源是传统互联网企业依旧在产生巨大的数据，如京东、淘宝，等等。例如淘宝双 11 的当天，交易额可以突破千亿元，由交易产生的数据高达 46GB，这仅仅是一天的数据。注意，这是顾客购物记录的文本数据，不是 46 集电视连续剧。

第二个来源是物联网的发展带来了大数据。我们已经进入了物联网时代，也就是在原来只有计算机组成的互联网基础上，加入许多非计算机节点。大街小巷的监控每天都在记录视频数据，物流中转站每天都在用手持设备扫描货物入库/出库记录，门禁数据，校园卡消费数据，家居智能产品产生的数据等。这些物联网设备在城市的每个角落随处可见，所以现在就有了智慧城市、智慧地球的概念，这是大数据很重要的一个来源。

第三个来源就是移动应用快速发展，人们都成了低头族，聊着 QQ，发着微信，顾不上跟人打招呼，都在忙着造大数据。移动应用也是大数据的重要来源。

1.3 如何处理大数据

随着数据越来越多，我们自然就会面临两个问题，第一个问题是这些大量历史数据还有用吗，为了节省空间可否删除呢？毕竟硬件资源有限。答案是有用的，而且可能带来意想不到的价值。我们可能从大量数据中找到某些行业的规律或规则，这些规则可能会带来巨大收益。第二个问题就是如何处理这些海量历史大数据呢？我们的处理办法就是传统的商业智能领域的数据挖掘技术。另外，还有一种处理技术是目前比较火的云计算技术，这种技术对数据处理的实时性要求很高，一般要求秒级处理。读者可以通过以下一个小例子初步了解一下数据挖掘技术和基于云平台的分布式处理技术。

1.3.1 数据分析与挖掘

下面分享一个真正通过数据挖掘收益的经典案例。最大零售超市沃尔玛拥有世界上最大的数据仓库系统。为了能够准确了解顾客在其门店的购买习惯，沃尔玛对其顾客的购物行为进行购物篮分析，想知道顾客经常一起购买的商品有哪些。通过数据挖掘和分析，一个意外的发现是：美国中年男子购买尿布的同时一般存在很大的可能会购买啤酒。通过分析后超市将啤酒和尿布摆放位置靠近，这样给超市带来了巨大收益。

国内百度大数据也做过很多次预测分析。2014 年世界杯足球赛，百度通过大数据分析了所有比赛，小组赛准确率达到 60%，淘汰赛阶段高达 100%。还有微软、Google 等几家公司都利用大数据做了相应的预测。如图 1-2 所示是几家公司预测对比。

	1/4决赛准确率	1/8决赛准确率	小组赛准确率
Baidu百度	100%	100%	58.33%
Microsoft	100%	100%	56.25%
Goldman Sachs	100%	100%	37.5%
Google	75%	100%	/

图 1-2

这两个案例都是传统的数据分析领域，最终目的都是为了从大数据中找到一些规则或者作出预测，为企业决策提供帮助，有点像沙里淘金。为了一点点金粒，就要留住所有沙子。这里的金子就是规则和结果，大量沙子就是大数据。

数据分析的步骤类似于从沙子里淘金的步骤，其步骤如下：

步骤 01 采集大数据（可能有很多来源，这里要说明一下，数据必须真实可靠，否则得到的规则也将是错误的）。

步骤 02 数据抽取（清洗，把对结果形成干扰的或者异常的数据剔除。比如运动员档案的数据里面出现一些名字，各项指标都是空着的，这样没意义的数据要删除）。

步骤 03 在清洗完毕的数据基础上构建数据仓库（实际上就是对我们感兴趣的维度构建一个模型，比如你要考察的是足球运动员，可能关注身高、体重、坐高、下肢长、小腿长，而对长得帅不帅、哪里人不感兴趣），模型建好之后，最后一步运用数据挖掘算法进行计算得到结论，这就是大数据处理的传统领域——数据分析，也叫作商业智能。

上面讲到的两个案例都是实时性要求不高，不要求马上得到结果。如果希望快速得到结果，比如几秒钟得到处理结果，这就是大数据处理的另一个领域，即云计算。本书不详细讲解云计算，此处仅仅举个小例子，以帮助读者理解云计算的概念。

1.3.2 基于云平台的分布式处理

介绍云计算之前，首先思考一个小问题，先不考虑会不会写代码，只要有想法即可。

问题：给出一篇文档，让你从中找出出现的单词以及这些单词出现的次数。

想想解题思路是怎样的？

解题思路其实很简单，就是从头到尾读取文档，碰到单词记录下来，同时记录它出现的次数，如果之前出现过，计数就加 1。具体可参考如图 1-3 所示的代码，有学过 Java 的读者可以参照这个代码实现方法理解其思路。

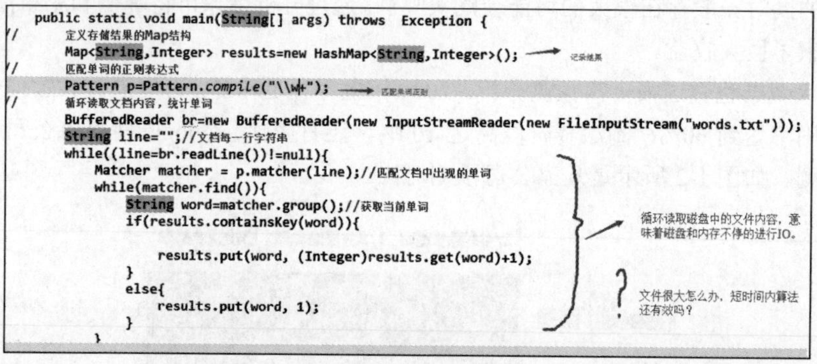

```java
public static void main(String[] args) throws Exception {
    //定义存储结果的Map结构
    Map<String,Integer> results=new HashMap<String,Integer>();    ←记录结果
    //匹配单词的正则表达式
    Pattern p=Pattern.compile("\\w+");    ←匹配单词压缩
    //循环读取文档内容，统计单词
    BufferedReader br=new BufferedReader(new InputStreamReader(new FileInputStream("words.txt")));
    String line="";//文档每一行字符串
    while((line=br.readLine())!=null){
        Matcher matcher = p.matcher(line);//匹配文档中出现的单词
        while(matcher.find()){
            String word=matcher.group();//获取当前单词
            if(results.containsKey(word)){
                results.put(word, (Integer)results.get(word)+1);
            }
            else{
                results.put(word, 1);
            }
        }
    }
}
```

循环读取磁盘中的文件内容，意味着磁盘和内存不停的进行IO。

文件很大怎么办，短时间内算法还有效吗？

图 1-3

理清了思路，再思考一个问题，如果读取一个 100GB 的文档，使用原来的算法还能处理吗？还能在合理的时间内给出答案吗？我们知道从硬盘读取文件到内存是通过 IO 流进行的，而计算的大量时间耗费都在 IO 上了。由于读取的数据体量很大，所以无论你对 WordCount 算法本身如何调整和优化，数据处理的效率依然会很低。那么应该如何快速处理呢？

有两种解决方法：一是找一台运算性能非常高的服务器，存储和运算能力都很惊人，但造价同样惊人。SAP 公司内存数据库产品 HANA，直接将所有数据存储到内存，全部数据存储和运算在内存中进行。但造价非常昂贵。国内有几家公司在使用这样的产品，比如农夫山泉。但并不是所有企业都负担得起。

另一个方法就是把一些廉价的服务器形成集群，每个服务器都需要承担一定运算任务，合作完成。就这个例子而言，可以用 100 台普通计算机，每台计算 1GB 的数据，最后统计出来的结果合并在一起就可以了。这个方法既节省成本，而且速度又快，因为服务器之间是并行运算的。这种技术早期叫网格计算，后来叫分布式计算，其实本质上就是如今的云计算。

上面这个思想需要解决两个问题：一是分布式存储，大数据分别存储到不同机器，而对使用者感觉好像是一台机器；二是分布式计算问题，每台机器都需要分配一个任务执行运算，所有任务同时进行，最后还需要对各个节点的运算结果进行合并得到结果。能够解决这两个问题流行的云计算框架就是 Apache 的 Hadoop 项目，里面包含好多的子项目和模块。Hadoop 官方网站页面如图 1-4 所示。

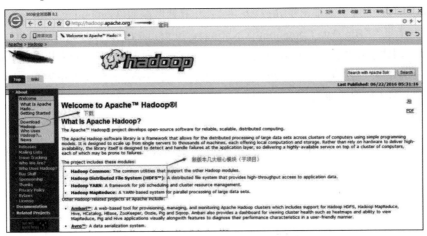

图 1-4

Hadoop 核心就是解决分布式存储的 HDFS 和解决分布式计算的 MapReduce。另外，它还有一些其他模块，也会涉及一些其他领域的技术，结合起来解决云计算问题。这些技术就像自然界的生态系统一样也构成了 Hadoop 生态圈，如图 1-5 所示。

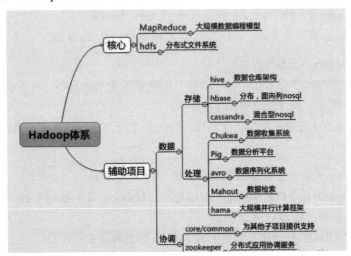

图 1-5

Hadoop 目前使用现状：Hadoop 被公认是一套行业大数据标准开源软件，在分布式环境下提供了海量数据的处理能力。几乎所有主流厂商都围绕 Hadoop 在开发工具、开源软件、商业化工具和技术服务。大型企业，如 Oracle、IBM、Microsoft、Intel、Cisco 都明显增加了 Hadoop 方面的投入。

淘宝从 2009 年开始，用于对海量数据的离线处理，例如对日志的分析、交易记录的分析等。规模从当初的 3~400 台节点，增加到现在的一个集群有 3000 个节点。淘宝现在已经有 2~3 个这样的集群。在支付宝的集群规模也有 700 台节点，使用 HBase 对用户的消费记录可以实现毫秒级查询。

随着以博客、社交网络、基于位置的服务 LBS 为代表的新型信息发布方式的不断涌现，以及云计算、物联网等技术的兴起，数据正以前所未有的速度在不断地增长和累积，大数据时代已经到来。学术界、工业界甚至于政府机构都已经开始密切关注大数据问题，并对其产生浓厚的兴趣。就学术界而言，Nature 早在 2008 年就推出了 BigData 专刊。计算社区联盟（Computing Community Consortium）在 2008 年发表了报告"Big Data Computing: Creating revolutionary breakthroughs in commernce, science, and society"，阐述了在数据驱动的研究背景下，解决大数据问题所需的技术以及面临的一些挑战。Science 期刊在 2011 年 2 月推出专刊 Dealing with Data，主要围绕着科学研究中大数据问题展开讨论，说明大数据对于科学研究的重要性。

1.4　Hadoop 3 新特性

根据官方网站 change log（修改日志）的介绍，Hadoop 3 中一些新增的特性介绍如下：

（1）最低支持 JDK1.8 及以上版本，不再支持 JDK1.7。Hadoop 版本与 JDK 版本之间的匹配关系为：

- Apache Hadoop 3.3 及以上版本支持 Java 8 和 Java 11（仅仅运行时）。请在编译 Hadoop 时使用 Java 8。Hadoop 在使用 Java 11 编译时不支持 HADOOP-16795，使用 Java 11 编译时支持 OPEN。
- Apache Hadoop 3.0.x 到 3.2.x 版本现在只支持 Java 8。
- Apache Hadoop 从 2.7.x 到 2.10.x 版本支持 Java 7 和 Java 8。

（2）YARN Timeline 版本升为 2.0。

（3）高可靠支持超过 2 个 NameNode 节点。如配置 3 个 NameNode 和 5 个 JournalNode。

（4）默认端口变化，具体变化如图 1-6 所示。

（5）从 Hadoop 2.9 开始添加了新的模块：Oozie，自此 Hadoop 拥有 5 个核心模块，以下是官方模块列表：

- Hadoop Common：支持其他 Hadoop 模块的常用工具。
- Hadoop Distributed File System（HDFS）：Hadoop 用于数据存储的分布式文件系统，提供应用数据的高吞吐量访问。
- Hadoop YARN：用于作业调度和集群资源管理的框架。
- Hadoop MapReduce：基于 YARN 框架，用于处理大数据集的分布式并行计算框架。
- Hadoop Ozone：是一个分布式对象存储系统，提供的是一个 key-value 形式的对象存储服务。

分类	应用	Haddop 2.x port	Haddop 3 port
NNPorts	Namenode	8020	9820
NNPorts	NN HTTP UI	50070	9870
NNPorts	NN HTTPS UI	50470	9871
SNN ports	SNN HTTP	50091	9869
SNN ports	SNN HTTP UI	50090	9868
DN ports	DN IPC	50020	9867
DN ports	DN	50010	9866
DN ports	DN HTTP UI	50075	9864
DN ports	Namenode	50475	9865

图 1-6

（6）Hadoop 3 之后，已经不再建议使用 root 用户启动和管理 Hadoop 的进程。建议创建一个非 root 用户来启动和管理 Hadoop 的进程。建议创建一个名称为 hadoop 的用户，并设置 hadoop 用户属于 wheel 组。

1.5 虚拟机与 Linux 操作系统的安装

1.5.1 VirtualBox 虚拟机的安装

本书将选择使用 VirtualBoxw 作为虚拟环境安装 Linux 和 Hadoop。VirtualBox 最早由 SUN 公司开发。由于 SUN 公司目前已经被 Oracle 收购，所以可以在 Oracle 公司的官方网站上下载到 VirtualBox 虚拟机软件的安装程序。下载地址为 https://www.virtualbox.org。到作者写作时，VirtualBox 的最新版本为 6.1.8。请在 VritualBox 的官方网站下载 Windows hosts 版本的 VirtualBox。其官方网站地址为 https://www.virtualbox.org/wiki/Downloads，如图 1-7 所示。

同时，由于 VitualBox 需要虚拟化 CPU 的支持，如果在安装虚拟机操作系统时，不支持安装 x64 位的 CentOS，可以在主机开机时按 F12 键进入宿主机的 BIOS 设置，并打开 CPU 的虚拟化设置。

打开 CPU 的虚拟化设置，界面如图 1-8 所示。

图 1-7　　　　　　　　　　　图 1-8

读者下载完成 VirtualBox 虚拟机后，自行安装即可。虚拟机的安装相对比较简单，以下是重要环节的部分截图。

网络组件安装如图 1-9 所示，单击"是"按钮。

图 1-9

网络组件安装如图 1-10 所示，单击"安装"按钮。

图 1-10

如图 1-9 所示，网络组件安装成功后，会在"我的网络"里面多出一个名为"Virtual Box Host Only"的本地网卡，此网卡用于宿主机与虚拟机通信。如图 1-11 所示。

图 1-11

1.5.2 Linux 操作系统的安装

本书将使用 CentOS7 作为虚拟机环境来学习和安装 Hadoop。首先需要下载 CentOS 操作系统，下载 minimal（最小）版本即可，因为我们使用 CentOS 并不需要可视化界面。然后启动 VirtualBox，如图 1-12 所示。

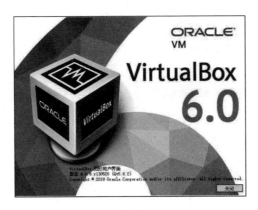

图 1-12

CentOS 的官方网站为 https://www.centos.org/。在官网上找到 CentOS Linux 安装包的下载页面，如图 1-13 所示，单击页面上的 CentOS-7-x86_64_minimal-2009.iso 链接，开始下载 minimal 版本。

Parent directory/	-	-
0_README.txt	2.4 KiB	2020-11-06 22:32
CentOS-7-x86_64-DVD-2009.iso	4.4 GiB	2020-11-04 19:37
CentOS-7-x86_64-DVD-2009.torrent	176.1 KiB	2020-11-06 22:44
CentOS-7-x86_64-Everything-2009.iso	9.5 GiB	2020-11-02 23:18
CentOS-7-x86_64-Everything-2009.torrent	380.6 KiB	2020-11-06 22:44
CentOS-7-x86_64-Minimal-2009.iso	973.0 MiB	2020-11-03 22:55
CentOS-7-x86_64-Minimal-2009.torrent	38.6 KiB	2020-11-06 22:44
CentOS-7-x86_64-NetInstall-2009.iso	575.0 MiB	2020-10-27 00:26
CentOS-7-x86_64-NetInstall-2009.torrent	23.0 KiB	2020-11-06 22:44
sha256sum.txt	398 B	2020-11-04 19:38
sha256sum.txt.asc	1.2 KiB	2020-11-06 22:37

图 1-13

下载完成后，将得到一个 CentOS-7-x86_64_minimal-2009.iso 文件。注意，文件名中的 2009 不是指 2009 年，而是指 2020 年 9 月发布的版本。

步骤 01 新建虚拟机，如图 1-14 所示。

步骤 02 输入操作系统的名称和选择操作系统的版本，如图 1-15 所示。

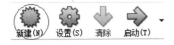

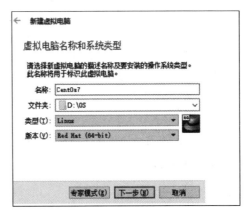

图 1-14

图 1-15

步骤 03 为新的系统分配内存，建议 4GB（最少 2GB）或以上，这要根据你自身主机的内存而定。同时建议设置 CPU 为 2 个。如图 1-16 和图 1-17 所示。

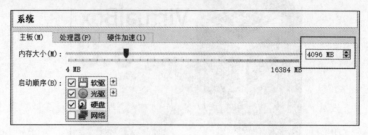

图 1-16

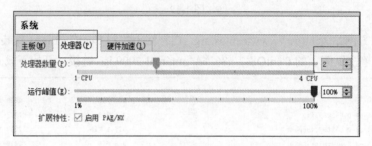

图 1-17

步骤 04 为新的系统创建硬盘，设置为动态增加，建议最大设置为 30GB 或以上。同时选择虚拟文件所保存的目录，默认情况下，会将虚拟化文件保存到 C:/盘上。笔者建议最好保存到非系统盘上，如 D:/OS 目录下将是不错的选择。如图 1-18 所示。

步骤 05 选择创建以后，右击进入设置界面，单击"存储"，在右侧的列表中选择已经下载的 CentOS7 的 iso 镜像文件。如图 1-19 所示。

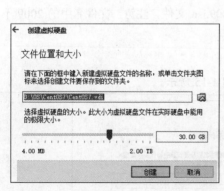

图 1-18

图 1-19

步骤 06 查看网络设置，将网卡 1 设置为 NAT 以用于连接外网，将网卡 2 设置为 Host Only 用于与宿主机进行通信。

网卡 1 的设置如图 1-20 所示。

图 1-20

网卡 2 的设置如图 1-21 所示。

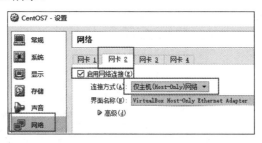

图 1-21

步骤 07 现在启动这个虚拟机，将会进入 CentOS7 的安装界面，选择 Install CentOS Linux 7，然后就开始安装 CentOS Linux，如图 1-22 所示。

图 1-22

步骤 08 在安装过程中出现选择语言项目，可以选择【中文】。选择安装，如图 1-23 所示。进入安装位置，选择整个磁盘即可，如图 1-24 所示。选择最小安装即可。注意，必须同时选择打开 CentOS 的网络，如图 1-25 所示。否则安装成功以后，CentOS 将没有网卡设置的选项。

图 1-23

图 1-24

图 1-25

步骤 09 在安装过程中，创建一个非 root 用户，并选择属于管理员组。在其后的操作中，笔者不建议使用 root 用于进行具体的操作。一般情况下，只要执行 sudo 即可以用 root 用户执行相关命令，输入的密码请牢记，如图 1-26 所示。

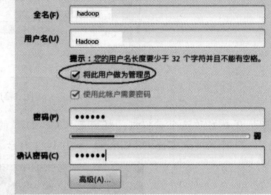

图 1-26

步骤 10 安装完成以后，重新启动虚拟机系统，并测试是否可以使用之前创建的用户名和密码登录。安装完成后的启动，请选择正常启动，正常启动即以有界面的方式启动，等我们设置好一些信息后，即可以选择无界面启动。启动方式选择有界面启动，如图 1-27 所示。

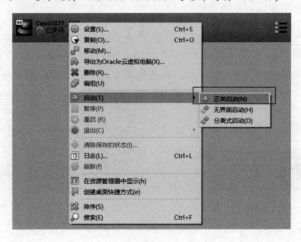

图 1-27

步骤⑪ 设置静态 IP 地址启动后，将显示如图 1-28 所示的界面，此时可以选择以 root 用户名和密码登录。注意输入密码时，将不会有任何的响应，此时不用担心，只要确认输入正确，按回车键后即可以看到登录成功后的界面，如图 1-29 所示。

图 1-28

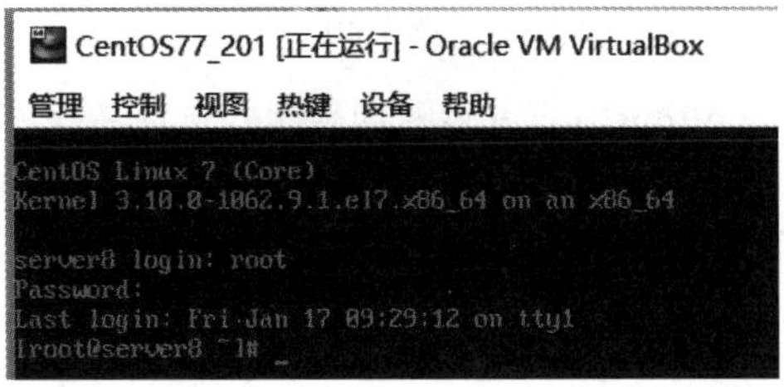

图 1-29

对于 Linux 系统来说，如果当前是 root 用户，将会显示#，如图 1-28 所示，root 用户登录成功后，将会显示[root@server8 ~]#，其中#表示当前为 root 用户。如果是非 root 用户，提示符将显示为$。

接下来设置静态 IP 地址。使用 vim 修改/etc/sysconfig/network-scripts/ifcfg-enp0s8，修改内容为：其中 IPADDR=192.168.56.201 为本 Linux 的 hostOnly 网卡地址，用于主机通信。输出完成以后，按 Esc 键，然后再输入:wq 保存退出即可。这是 vim 的基本操作，不了解的读者，可以去网上查看 vim 的基本使用。

```
TYPE=Ethernet
PROXY_METHOD=none
BROWSER_ONLY=no
BOOTPROTO=static
DEFROUTE=yes
IPV4_FAILURE_FATAL=no
IPV6INIT=yes
IPV6_AUTOCONF=yes
IPV6_DEFROUTE=yes
IPV6_FAILURE_FATAL=no
IPV6_ADDR_GEN_MODE=stable-privacy
NAME=enp0s8
UUID=620377da-1744-4268-b6d6-a519d27e01c6
DEVICE=enp0s8
ONBOOT=yes
IPADDR=192.168.56.201
```

请牢记上面设置的 IP 地址。现在可以关闭系统，并以无界面方式重新启动 CentOS。以后我们将使用 SSH 客户端登录此 CentOS。

上述文件是在配置了 HostOnly 网卡的情况下才会存在 ifcfg-enp0s8。如果没有这个文件，可关闭 Linux，并重新添加 HostOnly 网卡后，再进行配置。如果添加了 Hostonly 网卡后，依然没有此文件，可以在相同目录下，复制 ifcfg-enp0s3 为 ifcfg-enp0s8。

现在关闭 CentOS，以无界面方式启动，如图 1-30 所示。

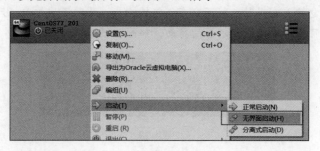

图 1-30

注意：① 本书不重点介绍 VirtualBox 虚拟机的使用，所以只给出具体的操作步骤。

② 在安装过程中，鼠标会在虚拟机和宿主机之间切换。如果要从虚拟机中退出鼠标请按 Ctrl 键即可。

③ 关于 Linux 命令请读者自行参考 Linux 手册。如：vim/vi、sudo、ls、cp、mv、tar、chmod、chown、scp、ssh-keygen、ssh-copy-id、cat、mkdir 等将是后面经常会使用到的命令。

1.6 SSH 工具与使用

读者可以选择 XShell、CRT、MobaXterm 等，作为 Linux 操作的远程命令行工具。同时配合它们的 xFtp 可以实现文件的上传与下载。

需要说明的是，XShell、CRT、MobaXterm 的文件上传方式各不相同。读者可以查看各软件相关文档实现上传，如图 1-31 所示为 MobaXterm 文件的上传方式。

图 1-31

XShell 和 CRT 是收费软件，不过读者可以在安装时，选择 free for school（学校免费版本）免费使用。安装完成以后，在命令行使用 SSH 即可以登录 Linux。MobaXterm 为免费软件，图 1-32 所示为 MobaXterm 通过 SSH 远程登录 Linux 的界面。

图 1-32

在弹出的界面上，选择 SSH，并输入主机名称和用户名登录，如图 1-33 所示。

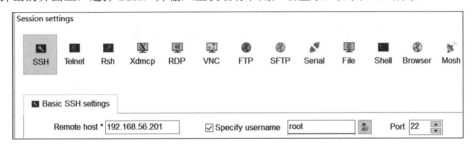

图 1-33

输入密码不会有任何的回显，只需要输入正确回车即可，如图 1-34 所示。

图 1-34

登录成功后的界面，如图 1-35 所示。

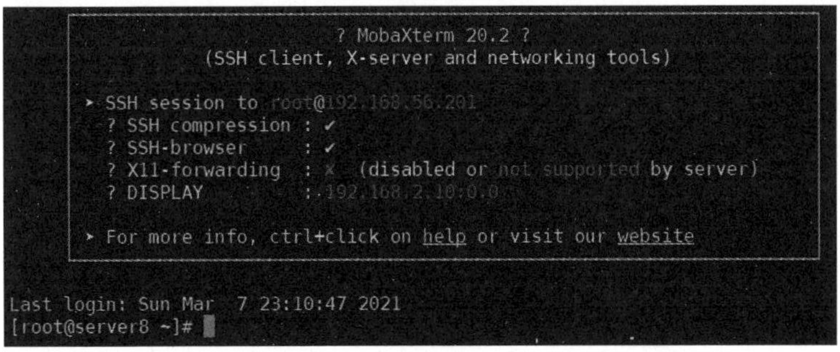

图 1-35

1.7 Linux 统一设置

在后面的章节中将使用一些 Linux 统一的设置，在此一并列出。由于本次登录（见图 1-29）是使用 root 登录的，可以直接操作某些命令，不用添加 sudo 命令。

步骤01 配置主机名称。

作为笔者的习惯，总喜欢将主机名取为 server+IP 最后一部分作为主机名称，下面例子取主机名为 server201，是因为本主机的 IP 地址设置为 192.168.56.201。

```
# hostnamectl dset-hostname server201
```

步骤02 修改 hosts 文件。

在 hosts 文件的最后，添加以下的配置，可以通过 # vim /etc/hosts 进行修改。

```
192.168.56.201    server201
```

步骤03 关闭且禁用防火墙。

```
# systemctl stop firewalld
# systemctl disable firewalld
```

步骤04 禁用 SELinux，需要重新启动。

```
#vim /etc/selinux/config
SELINUX=disabled
```

步骤05 设置时间同步（可选）。

```
#vim /etc/chrony.conf
```

删除所有 server，只添加：

```
server  ntp1.aliyun.com iburst
```

重新启动 chronyd：

```
#systemctl   restart chronyd
```

查看状态：

```
#chronyc       sources    -v
^* 120.25.115.20
```

如果显示*，表示时间同步成功。

步骤06 在 /usr/java 目录下，安装 JDK1.8.x。

usr 目录为 unix system resource，将 JDK1.8_x64 安装到此目录下。
首先在 Oracle 官网下载 JDK1.8 的 Linux 版本，如图 1-36 所示。

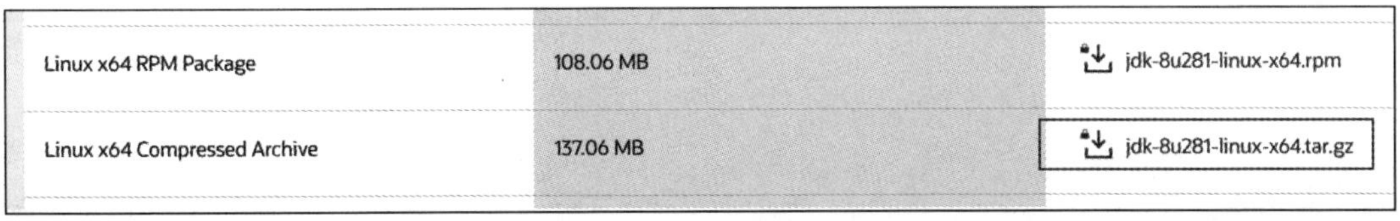

Linux x64 RPM Package	108.06 MB	jdk-8u281-linux-x64.rpm
Linux x64 Compressed Archive	137.06 MB	jdk-8u281-linux-x64.tar.gz

图 1-36

上传到 Linux 并解压：

`# tar -zxvf jdk-8u281-linux-x64.tar.gz -C /usr/java/`

步骤 07 配置 JAVA_HOME 环境变量。

`#vim /etc/profile`

在 profile 文件最后，添加以下配置：

```
export JAVA_HOME=/usr/java/jdk1.8.0_281
export PATH=.:$PATH:$JAVA_HOME/bin
```

让环境变量生效：

`# source /etc/profile`

检查 Java 版本：

```
[root@localhost bin]# java -version
java version "1.8.0_281"
Java(TM) SE Runtime Environment (build 1.8.0_192-b12)
Java HotSpot(TM) 64-Bit Server VM (build 25.192-b12, mixed mode)
```

至此，基本的环境就配置完成了。

1.8　本章小结

本章主要讲解了以下内容：

- 大数据概述。
- Hadoop 介绍。
- 虚拟机及网络的配置。
- CentOS7 Linux 操作系统的安装过程。
- 使用 XShell 登录 CentOS7。
- Linux 的基本命令。
- SSH 远程登录 Linux。
- 配置通用的一些配置。
- 安装 JDK 及配置环境变量。

第 2 章

Hadoop 伪分布式集群

本章主要内容：

- 安装独立运行的 Hadoop。
- Hadoop 伪分布式的安装与配置。
- HDFS 的命令。
- Java 操作 HDFS。

Hadoop 的运行方式可以分为 3 种：

- 独立运行的 Hadoop。不提供 HDFS 存储服务，也不需要启动任何的后台守护进程，但可以直接在本地运行 MapReduce 程序，并将输出结果保存到本地磁盘上。
- 伪分布式运行的 Hadoop。一般是指只有一台服务器的 Hadoop 运行环境，需要启动 NameNode（主节点存储服务）、SecondaryNameNode（主节点日志数据备份服务）可提供 HDFS 存储服务。启动守护进程 ResourceManager 和 NodeManager，运行 MapReduce 程序并将结果输出到 HDFS 上。
- 集群运行的 Hadoop。可用于生产环境的高可靠集群。借助 ZooKeeper 实现宕机容灾和自动切换。

为了快速上手，我们会运行一个独立的 MapReduce。独立运行的 MapReduce 可读取本地文本文件，然后将输出的数据保存到本地磁盘上。

注意：本书后面的环境，都使用 CentOS7、JDK1.8_x64 和 Hadoop 3.2.2 作为基础环境。本节搭建的伪分布式集群（实际上是单台虚拟机）所用的服务器及相关配置，可以用于第 2 章到第 12 章所有涉及的伪分布式操作环境。

2.1 安装独立运行的 Hadoop

独立运行的 Hadoop 可以帮助你快速运行一个 MapReduce 示例，以了解 MapReduce 的运行。后面的测试和基本命令将会运行在分布式环境下。有些应用，如 HBase、Hive 则需要真实的集群环境。

步骤 01 下载 Hadoop。Hadoop 3.2.2 的下载地址为 https://www.apache.org/dyn/closer.cgi/hadoop/common/hadoop-3.2.2/hadoop-3.2.2.tar.gz。

步骤 02 解压并配置环境。

以 hadoop 用户登录，并在/home/hadoop 的主目录下创建一个目录，用于安装 Hadoop。

```
$ mkdir ~/program
```

上传 Hadoop 压缩包，并解压到 program 目录下：

```
$ tar -zxvf hadoop-3.1.3.tar.gz -C ~/program/
```

配置 Java 的环境变量，修改 hadoop 解压目录下的/etc/hadoop/hadoop-env.sh 文件，找到${JAVA_HOME}配置项并设置为本机 JAVA_HOME 的地址。

```
$ vim ~/program/hadoop-3.2.2/etc/hadoop/hadoop-env.sh
export JAVA_HOME=/usr/java/jdk1.8.0_281
```

配置 Hadoop 环境变量：

```
$ vim /home/hadoop/.bash_profile
export HADOOP_HOME=/home/hadoop/program/hadoop-3.2.2
export PATH=$PATH:$HADOOP_HOME/bin
```

注意：由于笔者是用 hadoop 用户登录系统的，只配置了 hadoop 用户的环境变量，在这种情况下，这种配置只能当前用户可用。读者可以根据自己的要求进行配置。如：如果配置到/etc/profile 文件中，则是整个系统都可以使用的环境变量，那么就不需要将 Hadoop 安装到某个用户的主目录下了。

让环境变量生效：

```
$source ~/.bash_profile
```

输入 hadoop 命令，查看 Hadoop 的版本：

```
[hadoop@server201 ~]$ hadoop version
Hadoop 3.2.2
Source code repository Unknown -r 7a3bc90b05f257c8ace2f76d74264906f0f7a932
Compiled on 2021-01-03T09:26Z
Compiled with protoc 2.5.0
From source with checksum 5a8f564f46624254b27f6a33126ff4
This command was run using /home/hadoop/program/hadoop-3.2.2/share/hadoop/common/hadoop-common-3.2.2.jar
```

步骤 03 独立运行 MapReduce。

Hadoop 可以运行在一个非分布式的环境下,即可以运行为一个独立的 Java 进程。现在运行一个 wordcount 的 MapReduce 示例。

创建一个任意的文本文件,并输入一行英文单词:

```
[hadoop@server201 ~]$ touch a.txt
[hadoop@server201 ~]$ vim a.txt
Hello This is
a Very Sample MapReduce
Example of Word Count
Hope You Run This Program Success!
```

执行 wordcount 测试:

```
[hadoop@server201 ~]$ hadoop jar \
~/program/hadoop-3.2.2/share/hadoop/mapreduce/hadoop-mapreduce-examples-3.2.2.jar \
 wordcount \
~/a.txt \
~/out
```

命令执行成功后会显示以下信息,注意输出的日志会比较多,请仔细查找。

```
2021-03-08 21:59:19,536 INFO mapreduce.Job:  map 100% reduce 100%
2021-03-08 21:59:19,537 INFO mapreduce.Job: Job job_local215774179_0001 completed successfully
```

命令说明:

- **hadoop jar**:用于执行一个 MapReduce 示例。在 Linux 中,如果命令有多行,可以通过输入"\"(斜线)换行。注意"\"前面必须有空格。
- **hadoop-mapreduce-examples-3.1.3.jar**:为官方提供的示例程序包,wordcount 是执行的任务,~/a.txt 是输入的目录或文件,~/out 是程序执行成功以后的输出目录。

程序执行成功以后,进入 out 输出目录,查看输出目录中的数据文件,其中 part-r-0000 为数据文件,_SUCCESS 为标识成功的文件,其中没有数据。

```
[hadoop@server201 ~]$ cd out/
[hadoop@server201 out]$ ll
总用量 4
-rw-r--r-- 1 hadoop hadoop 122 3月   8 21:59 part-r-00000
-rw-r--r-- 1 hadoop hadoop   0 3月   8 21:59 _SUCCESS
```

通过 cat 查看 part-r-00000 文件中的数据,可以看到已经对 a.txt 中的单词进行了数量统计,且默认排序为字母的顺序,字母后是此单词出现的次数。

```
[hadoop@server201 out]$ cat *
Count    1
Example  1
Hello    1
Hope     1
MapReduce        1
```

```
Program     1
Run         1
Sample      1
Success!    1
This        2
Very        1
Word        1
You         1
a           1
is          1
of          1
```

可见，已经对<input>目录中文件的数据进行统计。至此，独立运行模式的 Hadoop 已经搭建完成。Hadoop 独立运行方式只是一个练习，在正式的运行环境中不会使用这种方式。这里只是让大家了解一下 MapReduce 的运行。而且在此模式下，Hadoop 的 HDFS 不会运行，也不会存储数据。

2.2 Hadoop 伪分布式环境准备

Hadopp 伪分布式，即在单机模式下运行 Hadoop。我们需要运行 5 个守护进程，其中 3 个负责 HDFS 存储，2 个负责 MapReduce 计算。

负责 HDFS 存储的 3 个进程如下（见图 2-1）：

- NameNode 进程：作为主节点，主要负责分配数据存储的位置。
- SecondaryNameNode 进程：作为 NameNode 日志备份和恢复进程，避免数据丢失。
- DataNode 进程：作为数据的存储节点，接收客户端的数据读写请求。

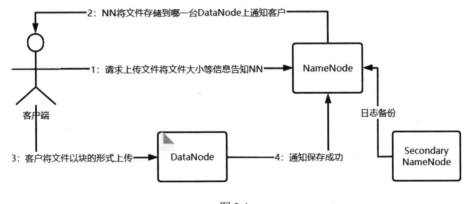

图 2-1

负责 MapReduce 计算的 2 个进程如下：

- ResourceManager 进程：负责分配计算任务由哪一台主机执行。
- NodeManager 进程：负责执行计算任务。

在真实集群环境下，部署的一般规则是：

- 由于 NodeManger 需要读取 DataNode 上的数据，用于执行计算，所以一般 DataNode 与 NodeManger 并存。
- 由于 NameNode 在运行时，需要在内存中大量缓存文件块的数据，所以 NameNode 节点都应该部署到内存比较大的主机上。
- 在真实的集群环境下，一般部署多个 NameNode 节点，互为备份和切换关系。且不再部署 SecondaryNameNode 进程。

伪分布式可以让读者快速学习 HDFS 的命令及开发 MapReduce 应用，对于学习 Hadoop 有很大的帮助。在安装之前，笔者有以下建议：

- 配置静态 IP 地址。虽然是单机模式，但也建议配置静态的 IP 地址，这有助于以后配置集群环境时固定 IP，养成良好的习惯。
- 修改主机名称为一个便于记忆的名称，如 server201，修改规则一般为本机的 IP 地址最后一段作为服务器的后缀，如：192.168.56.201，则可以修改本主机的名称为 server201。
- 由于启动 Hadoop 的各个进程使用的是 SSH。所以，必须配置本机免密码登录。本章后面的步骤会介绍如何配置 SSH 免密码登录。配置 SSH 免密码登录的规则是在启动的集群的主机上，向其他主机配置 SSH 免密码登录，以便于操作机可以在不登录其他主机的情况下，启动所需要的进程。
- 关闭防火墙。如果你的 CentOS7 没有安装防火墙，可以不用关闭，如果已经安装了，请检查防火墙的状态，如果是运行状态，请关闭防火墙并禁用防火墙。注意，在生产环境下，不要直接禁用防火墙，可以指定 Hadoop 的某些端口开放。
- 使用非 root 用户，前面章节我们创建了一个名为 hadoop 用户，此用户同时属于 wheel 组（拥有此组的用户可以使用 sudo 命令，执行一些 root 用户的操作）。我们就以此用户作为执行命令的用户。

步骤 01 配置静态 IP 地址。

前面的章节已经讲过静态 IP 的设置，此处再做详细讲解。使用 SSH 登录 CentOS7。然后使用 ifconfig 查看 IP 地址，如果没有 ifconfig 命令，可以使用 sudo yum -y install net-tools 安装 ifconfig 命令。其实在 CentOS7 中，已经使用 ip addr 命令显示当前主机的 IP 地址。所以，也可以不安装 net-tools。

```
$ ifconfig
enp0s3: flags=4163<UP,BROADCAST,RUNNING,MULTICAST>  mtu 1500
        inet 10.0.2.15  netmask 255.255.255.0  broadcast 10.0.2.255
enp0s8: flags=4163<UP,BROADCAST,RUNNING,MULTICAST>  mtu 1500
        inet 192.168.56.201  netmask 255.255.255.0
```

上例显示为两块网卡，其中 enp0s3 的 IP 地址为 10.0.2.15，此网卡为 NAT 网络，用于上外网。enp0s8 的 IP 地址为 192.168.56.201，此网卡为 Host Only 网络，用于与宿主机进行通信。我们要修改的就是 enp0s8 这个网卡，将它的 IP 地址设置为固定 IP。

IP 设置保存在文件中，这个文件为/etc/sysconfig/network-scripts/ifcfg-enp0s8。使用 cd 命令，切换到这个目录下。使用 ls 命令显示这个目录下的所有文件，你可能只会发现 ifcfg-enp0s3 这个文件，现在使用 cp 命令将 ifcfg-enp0s3 复制一份为 ifcfg-enp0s8。由于 etc 目录不属于 hadoop 用户，所以操作时,需要添加 sudo 前缀。

```
$ sudo cp ifcfg-enp0s3 ifcfg-enp0s8
```

使用 vim 命令修改为静态 IP 地址:

```
$ sudo vim ifcfg-enp0s8
```

将原来的 dhcp 修改成 static 即静态的地址,并设置具体的 IP 地址。其中,每一个网卡,都应该具有唯一的 UUID,所以建议修改其中任意一个值,以便于与之前的 enp0s3 的 UUID 不同。部分修改内容如下:

```
BOOTPROTO="static"
NAME="enp0s8"
UUID="d2a8bd92-cf0d-4471-8967-3c8aee78d101"
DEVICE="enp0s8"
IPADDR="192.168.56.201"
```

现在重新启动网络:

```
$ sudo systemctl restart network.service
```

重新启动网络后,再次查看 IP,地址已经发生变化:

```
[hadoop@server201 ~]$ ifconfig
enp0s3: flags=4163<UP,BROADCAST,RUNNING,MULTICAST>  mtu 1500
        inet 10.0.2.15  netmask 255.255.255.0  broadcast 10.0.2.255
enp0s8: flags=4163<UP,BROADCAST,RUNNING,MULTICAST>  mtu 1500
        inet 192.168.56.201  netmask 255.255.255.0  broadcast 0x20<link>
lo: flags=73<UP,LOOPBACK,RUNNING>  mtu 65536
        inet 127.0.0.1  netmask 255.0.0.0
```

步骤 02 修改主机名称。

使用 hostname 命令,检查当前主机的名称:

```
$ hostname
localhost
```

使用 hostnamectl 命令,修改主机的名称:

```
$ sudo hostnamectl set-hostname server201
```

步骤 03 配置 hosts 文件。

hosts 文件是本地 DNS 解析文件。配置此文件,可以根据主机名找到对应的 IP 地址。

使用 vi 命令,打开这个文件,并在文件中追加以下配置:

```
$ sudo vim /etc/hosts
192.168.56.201  server201
```

步骤 04 关闭防火墙。

CentOS7 默认情况下没有安装防火墙。可以通过命令 sudo firewall-cmd --state 检查防火墙的状态,如果显示"command not found"一般为没有安装防火墙,此步可以忽略。以下命令检查防火墙

的状态:

```
$ sudo firewall-cmd --state
running
```

running 表示防火墙正在运行。以下命令用于停止和禁用防火墙:

```
$ sudo systemctl stop firewalld.service
$ sudo systemctl disable firewalld.service
```

步骤 05 配置免密码登录。

配置免密码登录的主要目的,就是在使用 Hadoop 脚本启动 Hadoop 的守护进程时,不需要再提示用户输入密码。SSH 免密码登录的主要实现机制,就是在本地生成一个公钥,然后将公钥配置到需要被免密码登录的主机上,登录时自己持有私钥与公钥进行匹配,如果匹配成功,则登录成功,否则登录失败。

可以使用 ssh-keygen 命令生成公钥和私钥文件,并将公钥文件复制到被 SSH 登录的主机上。以下是 ssh-keygen 命令,输入后直接按两次回车即可生成公钥和私钥文件:

```
[hadoop@server201 ~]$ ssh-keygen -t rsa
Generating public/private rsa key pair.
Enter file in which to save the key (/home/hadoop/.ssh/id_rsa):
Created directory '/home/hadoop/.ssh'.
Enter passphrase (empty for no passphrase):
Enter same passphrase again:
Your identification has been saved in /home/hadoop/.ssh/id_rsa.
Your public key has been saved in /home/hadoop/.ssh/id_rsa.pub.
The key fingerprint is:
SHA256:IDIO32gBEDXhFVE1l6oYca5P4fkfIZRywyhgJ4Id/I4 hadoop@server201
The key's randomart image is:
+---[RSA 2048]----+
|=*%+*+..o ..     |
|.=o0.+.o +.      |
| +.*+= *.        |
|  +ooo=..        |
|   o = +S .      |
|   E + =         |
|    o..          |
|    ...          |
|     ..          |
+-----[SHA256]-----+
```

如上面所说,生成的公钥和私钥文件将被放到~/.ssh/目录下。其中 **id_rsa** 文件为私钥文件,**rd_rsa.pub** 为公钥文件。现在我们再使用 ssh-copy-id 将公钥文件发送到目标主机。由于登录的是本机,所以直接输入本机名即可:

```
[hadoop@server201 ~]$ ssh-copy-id server201
/usr/bin/ssh-copy-id: INFO: Source of key(s) to be installed: "/home/hadoop/.ssh/id_rsa.pub"
The authenticity of host 'server201 (192.168.56.201)' can't be established.
ECDSA key fingerprint is SHA256:KqSRs/H1WxHrBF/tfM67PeiqqcRZuK4ooAr+xT5Z4OI.
```

```
    ECDSA key fingerprint is MD5:05:04:dc:d4:ed:ed:68:1c:49:62:7f:1b:19:63:5d:8
e.
    Are you sure you want to continue connecting (yes/no)? yes 输入 yes
    /usr/bin/ssh-copy-id: INFO: attempting to log in with the new key(s), to filt
er out any that are already installed
    /usr/bin/ssh-copy-id: INFO: 1 key(s) remain to be installed -- if you are pro
mpted now it is to install the new keys
```

输入密码然后按回车键，将会提示成功信息：

```
hadoop@server201's password:
Number of key(s) added: 1
Now try logging into the machine, with:   "ssh 'server201'"
and check to make sure that only the key(s) you wanted were added.
```

此命令执行以后，会在~/.ssh 目录下多出一个用于认证的文件，其中保存了某个主机可以登录的公钥信息，这个文件为~/.ssh/authorized_keys。如果读者感兴趣，可以使用 cat 命令查看这个文件中的内容。此文件中的内容，就是 id_rsa.pub 文件中的内容。

现在再使用 ssh scrvcr201 命令登录本机，将会发现不用再输入密码，即可以直接登录成功。

```
[hadoop@server201 ~]$ ssh server201
Last login: Tue Mar  9 20:52:56 2021 from 192.168.56.1
```

2.3 Hadoop 伪分布式安装

经过上面的环境设置，我们已经可以正式安装 Hadoop 伪分布式了。在安装之前，请确定你已经安装了 JDK1.8，并正确配置了 JAVA_HOME、PATH 环境变量。

在磁盘根目录下，创建一个 app 目录，并授权给 hadoop 用户。我们将会把 Hadoop 安装到此目录下。先切换到根目录下：

```
[hadoop@server201 ~]$ cd /
```

添加 sudo 前缀使用 mkdir 创建/app 目录：

```
[hadoop@server201 /]$ sudo mkdir /app
```

将此目录的所有权授予给 hadoop 用户和 hadoop 组：

```
[hadoop@server201 /]$ sudo chown hadoop:hadoop /app
```

切换进入/app 目录：

```
[hadoop@server201 /]$ cd /app/
```

使用 ll -d 命令查看本目录的详细信息，可看到此目录已经属于 hadoop 用户：

```
[hadoop@server201 app]$ ll -d
drwxr-xr-x 2 hadoop hadoop 6 3月   9 21:35 .
```

将 Hadoop 的压缩包上传到/app 目录下，并解压。

使用 ll 命令查看本目录，可以看到已经存在 hadoop-3.2.2.tar.gz 文件：

```
[hadoop@server201 app]$ ll
总用量 386184
-rw-rw-r-- 1 hadoop hadoop 395448622 3月   9 21:40 hadoop-3.2.2.tar.gz
```

使用 tar 命令 -zxvf 参数解压此文件：

```
[hadoop@server201 app]$ tar -zxvf hadoop-3.2.2.tar.gz
```

解压后，将看到一个目录 hadoop-3.2.2，此时可以删除 tar.gz 文件，此文件已经不再需要。

查看 /app 目录下，已经多出 hadoop-3.2.2 目录：

```
[hadoop@server201 app]$ ll
总用量 386184
drwxr-xr-x 9 hadoop hadoop       149 1月   3 18:11 hadoop-3.2.2
-rw-rw-r-- 1 hadoop hadoop 395448622 3月   9 21:40 hadoop-3.2.2.tar.gz
```

删除 hadoop-3.2.2.tar.gz 文件，此文件已经不再需要：

```
[hadoop@server201 app]$ rm -rf hadoop-3.2.2.tar.gz
```

以下开始配置 Hadoop。Hadoop 的所有配置文件都在 hadoop-3.2.2/etc/hadoop 目录下。首先切换到此目录下，然后开始配置：

```
[hadoop@server201 hadoop-3.2.2]$ cd /app/hadoop-3.2.2/etc/hadoop/
```

在 Hadoop 的官网上，有关于伪分布式配置的完成教程，地址为 https://hadoop.apache.org/docs/stable/hadoop-project-dist/hadoop-common/SingleCluster.html#Configuration。

大家也可以根据此教程进行伪分布式的配置学习。

步骤 01 配置 hadoop-env.sh 文件。

hadoop-env.sh 文件是 Hadoop 的环境文件，在此文件中需要配置 JAVA_HOME 变量。在此文件的第 55 行输入以下配置，然后按 Esc 键，再输入 :wq 保存退出即可：

```
export JAVA_HOME=/usr/java/jdk1.8.0_281
```

步骤 02 配置 core-site.xml 文件。

core-site.xml 文件是 HDFS 的核心配置文件，用于配置 HDFS 的协议、端口号和地址。

注意：Hadoop 3.0 以后 HDFS 的端口号建议为 8020，但如果查看 Hadoop 的官网示例，依然延续使用的是 Hadoop 2 之前的端口 9000，以下配置我们将使用 8020 端口，只要保证配置的端口没有被占用即可。配置时，需要注意大小写。

使用 vim 打开 core-site.xml 文件，进入编辑模式：

```
[hadoop@server201 hadoop]$ vim core-site.xml
```

在 <configuration></configuration> 两个标签之间输入以下内容：

```
<property>
   <name>fs.defaultFS</name>
   <value>hdfs://server201:8020</value>
```

```
    </property>
    <property>
        <name>hadoop.tmp.dir</name>
        <value>/opt/datas/hadoop</value>
    </property>
```

配置说明：

- fs.defaultFS：用于配置 HDFS 的主协议，默认为 file:///。
- hadoop.tmp.dir：用于指定 NameNode 日志及数据的存储目录，默认为/tmp。

步骤 03 配置 hdfs-site.xml 文件。

hdfs-site.xml 文件用于配置 HDFS 的存储信息。使用 vim 打开 hdfs-site.xml 文件，并在 <configuration></configuration>标签中输入以下内容：

```
    <property>
        <name>dfs.replication</name>
        <value>1</value>
    </property>
    <property>
        <name>dfs.permissions.enabled</name>
        <value>false</value>
    </property>
```

配置说明：

- dfs.replication：用于指定文件块的副本数量。HDFS 特别适合于存储大文件，它会将大文件切分成每 128MB 一块，存储到不同的 DataNode 节点上，且默认会每一块备份 2 份，共 3 份，即此配置的默认值为 3，最大为 512。由于我们只有一个 DataNode，所以这儿将文件副本数量修改为 1。
- dfs.permissions.enabled：访问时，是否检查安全，默认为 true。为了方便访问，暂时把它修改为 false。

步骤 04 配置 mapred-site.xml 文件。

通过名称可见，此文件是用于配置 MapReduce 的配置文件。通过 vim 打开此文件，并在 <configuration>标签中输入以下配置：

```
$ vim mapred-site.xml
    <property>
        <name>mapreduce.framework.name</name>
        <value>yarn</value>
    </property>
```

配置说明：

- mapreduce.framework.name：用于指定调试方式。这里指定使用 YARN 作为任务调用方式。

步骤 05 配置 yarn-site.xml 文件。

由于上面指定了使用 YARN 作为任务调度，所以这里需要配置 YARN 的配置信息，同样，使用 vim 编辑 yarn-site.xml 文件，并在<configuration>标签中输入以下内容：

```
<property>
    <name>yarn.resourcemanager.hostname</name>
    <value>server201</value>
</property>
<property>
    <name>yarn.nodemanager.aux-services</name>
    <value>mapreduce_shuffle</value>
</property>
```

通过 hadoop classpath 命令获取所有 classpath 的目录，然后配置到上述文件中。

由于没有配置 Hadoop 的环境变量，所以这里需要输入完整的 Hadoop 运行目录，命令如下：

```
[hadoop@server201 hadoop]$ /app/hadoop-3.2.2/bin/hadoop classpath
```

输入完成后，将显示所有 classpath 信息：

```
/home/hadoop/program/hadoop-3.2.2/etc/hadoop:/home/hadoop/program/hadoop-3.2.2/share/hadoop/common/lib/*:/home/hadoop/program/hadoop-3.2.2/share/hadoop/common/*:/home/hadoop/program/hadoop-3.2.2/share/hadoop/hdfs:/home/hadoop/program/hadoop-3.2.2/share/hadoop/hdfs/lib/*:/home/hadoop/program/hadoop-3.2.2/share/hadoop/hdfs/*:/home/hadoop/program/hadoop-3.2.2/share/hadoop/mapreduce/lib/*:/home/hadoop/program/hadoop-3.2.2/share/hadoop/mapreduce/*:/home/hadoop/program/hadoop-3.2.2/share/hadoop/yarn:/home/hadoop/program/hadoop-3.2.2/share/hadoop/yarn/lib/*:/home/hadoop/program/hadoop-3.2.2/share/hadoop/yarn/*
```

然后将上述的信息复制一下，并配置到 yarn-site.xml 文件中：

```
<property>
    <name>yarn.application.classpath</name>
    <value>
/home/hadoop/program/hadoop-3.2.2/etc/hadoop:/home/hadoop/program/hadoop-3.2.2/share/hadoop/common/lib/*:/home/hadoop/program/hadoop-3.2.2/share/hadoop/common/*:/home/hadoop/program/hadoop-3.2.2/share/hadoop/hdfs:/home/hadoop/program/hadoop-3.2.2/share/hadoop/hdfs/lib/*:/home/hadoop/program/hadoop-3.2.2/share/hadoop/hdfs/*:/home/hadoop/program/hadoop-3.2.2/share/hadoop/mapreduce/lib/*:/home/hadoop/program/hadoop-3.2.2/share/hadoop/mapreduce/*:/home/hadoop/program/hadoop-3.2.2/share/hadoop/yarn:/home/hadoop/program/hadoop-3.2.2/share/hadoop/yarn/lib/*:/home/hadoop/program/hadoop-3.2.2/share/hadoop/yarn/*
    </value>
</property>
```

配置说明：

- yarn.resourcemanager.hostname：用于指定 ResourceManger 的运行主机，默认为 0.0.0.0，即本机。
- yarn.nodemanager.aux-services：用于指定执行计算的方式为 mapreduce_shuffle。
- yarn.application.classpath：用于指定运算时的类加载目录。

步骤06 配置 workers 文件。

这个文件在之前的版本叫作 slaves，但功能一样。主要用于在启动时启动 DataNode 和 NodeManager。

编辑 workers 文件，并输入本地名称：

```
server201
```

步骤 07 配置 Hadoop 环境变量。

编辑/etc/profile 文件：

```
$ sudo vim /etc/profile
```

并在里面添加以下内容：

```
export HADOOP_HOME=/app/hadoop-3.2.2
export PATH=$PATH:$HADOOP_HOME/bin
```

使用 source 命令，让环境变量生效：

```
$ source /etc/profile
```

然后使用 hdfs version 查看命令环境变量是否生效，如果配置成功，则会显示 Hadoop 的版本：

```
[hadoop@server201 hadoop]$ hdfs version
Hadoop 3.2.2
Source code repository Unknown -r 7a3bc90b05f257c8ace2f76d74264906f0f7a932
Compiled by hexiaoqiao on 2021-01-03T09:26Z
Compiled with protoc 2.5.0
From source with checksum 5a8f564f46624254b27f6a33126ff4
This command was run using /app/hadoop-3.2.2/share/hadoop/common/hadoop-common-3.2.2.jar
```

步骤 08 初始化 Hadoop 的文件系统。

Hadoop 在使用之前，必须先初始化 HDFS 文件系统，初始化的文件系统将会生成在 hadoop.tmp.dir 配置的目录下，即上面配置的/app/datas/hadoop 目录下。

```
$ hdfs namenode -format
```

在执行命令完成以后，请在输出的日志中找到以下这句话，即为初始化成功：

```
Storage directory /opt/hadoop_tmp_dir/dfs/name has been successfully formatted.
```

步骤 09 启动 HDFS 和 YARN。

启动和停止 HDFS 及 YARN 的脚本在$HADOOP_HOME/sbin 目录下。其中 start-dfs.sh 为启动 HDFS 的脚本，start-yarn.sh 为启动 ResourceManager 的脚本。以下命令分别启动 HDFS 和 YARN：

```
[hadoop@server201 /]$ /app/hadoop-3.2.2/sbin/start-dfs.sh
[hadoop@server201 /]$ /app/hadoop-3.2.2/sbin/start-yarn.sh
```

启动完成以后，通过 jps 来查看 Java 进程快照，你会发现有 5 个进程正在运行：

```
[hadoop@server201 /]$ jps
```

```
12369 NodeManager
12247 ResourceManager
11704 NameNode
12025 SecondaryNameNode
12686 Jps
11839 DataNode
```

其中：NameNode、SecondaryNameNode、DataNode 是通过 start-dfs.sh 脚本启动的。ResourceManager 和 NodeManager 是通过 start-yarn.sh 脚本启动的。

启动成功以后，也可以通过 http://server201:9870 查看 NameNode 的信息，如图 2-2 所示。

图 2-2

可以通过 http://server201:8088 页面查看 MapReduce 的信息，如图 2-3 所示。

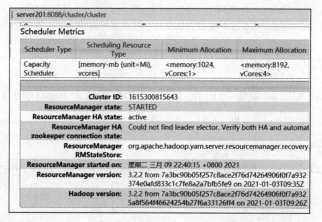

图 2-3

步骤 10 关闭。

关闭 HDFS 的 YARN 可以分别执行 stop-dfs.sh 和 stop-yarn.sh 脚本：

```
[hadoop@server201 /]$ /app/hadoop-3.2.2/sbin/stop-yarn.sh
Stopping nodemanagers
Stopping resourcemanager
[hadoop@server201 /]$ /app/hadoop-3.2.2/sbin/stop-dfs.sh
Stopping namenodes on [server201]
Stopping datanodes
```

```
Stopping secondary namenodes [server201]
```

至此，Hadoop 单机即伪分布式模式安装和配置成功。下一节将学习使用 Hadoop 的 hdfs 命令操作 Hadoop 的 HDFS 文件系统。

2.4 HDFS 操作命令

hdfs 命令位于 $HADOOP_HOME/bin 目录下。由于已经配置了 HADOOP_HOME 和 PATH 的环境变量，所以此命令可以在任意目录下执行。可以通过直接输入 hdfs 命令，查看它的使用帮助：

```
$ hdfs
Usage: hdfs [--config confdir] [--loglevel loglevel] COMMAND
       where COMMAND is one of:
  dfs                  run a filesystem command on the file systems supported in Hadoop.
  classpath            prints the classpath
  namenode -format     format the DFS filesystem
  secondarynamenode    run the DFS secondary namenode
  namenode             run the DFS namenode
  journalnode          run the DFS journalnode
  zkfc                 run the ZK Failover Controller daemon
  datanode             run a DFS datanode
  debug                run a Debug Admin to execute HDFS debug commands
  dfsadmin             run a DFS admin client
  haadmin              run a DFS HA admin client
  fsck                 run a DFS filesystem checking utility
  balancer             run a cluster balancing utility
  jmxget               get JMX exported values from NameNode or DataNode.
  mover                run a utility to move block replicas across
                       storage types
  oiv                  apply the offline fsimage viewer to an fsimage
  oiv_legacy           apply the offline fsimage viewer to an legacy fsimage
  oev                  apply the offline edits viewer to an edits file
  fetchdt              fetch a delegation token from the NameNode
  getconf              get config values from configuration
  groups               get the groups which users belong to
  snapshotDiff         diff two snapshots of a directory or diff the
                       current directory contents with a snapshot
  lsSnapshottableDir   list all snapshottable dirs owned by the current user
                                Use -help to see options
  portmap              run a portmap service
  nfs3                 run an NFS version 3 gateway
  cacheadmin           configure the HDFS cache
  crypto               configure HDFS encryption zones
  storagepolicies      list/get/set block storage policies
  version              print the version
Most commands print help when invoked w/o parameters.
```

上面的这些命令，在后面的课程中基本都会涉及。现在让我们来查看几个使用比较多的命令。

在上面的列表中,第一个 dfs 是经常被使用的命令。可以通过 hdfs dfs -help 查看 dfs 的具体使用方法。由于参数过多,本书就不一一列举了。dfs 命令,就是通过命令行操作 HDFS 目录或是文件的命令,类似于 Linux 文件命令一样,只不过 dfs 操作的是 HDFS 文件系统中的文件。表 2-1 列出 HDFS 几个常用命令。

表 2-1 HDFS 常用命令

命 令	功 能	示 例
-ls	用于显示 HDFS 文件系统上的所有目录和文件	hdfs dfs -ls /
-mkdir	在 HDFS 上创建一个新的目录	hdfs dfs -mkdir /test
-rm -r	删除 HDFS 上的一个目录,其中-r 参数为递归删除所有子目录。如果没有使用-r 参数,则是删除一个文件	hdfs dfs -rm -r /test
-put	将本地文件上传到 HDFS 上去	hdfs dfs -put a.txt /test/a.txt
-cat	显示 HDFS 上某个文件中的所有数据,如果给出的是一个目录,则会忽略目录以下"."(点)或是"_"(下画线)开始的文件	hdfs dfs -cat /test/a.txt
-get	从 HDFS 上将文件保存到本地	hdfs dfs -get /test/a.txt b.txt
-moveFromLocal	将本地文件上传到 HDFS 后并删除本地文件	hdfs dfs -moveFromLocal a.txt /test/a1.txt

下面看几个示例。显示根目录下的所有文件和目录:

```
[hadoop@server201 ~]$ hdfs dfs -ls /
Found 1 items
drwxr-xr-x   - hadoop supergroup          0 2021-03-10 20:41 /test
```

以递归的形式显示根目录下的所有文件或目录,注意-R 参数:

```
[hadoop@server201 ~]$ hdfs dfs -ls -R /
drwxr-xr-x   - hadoop supergroup          0 2021-03-10 20:41 /test
-rw-r--r--   1 hadoop supergroup          6 2021-03-10 20:41 /test/a.txt
```

删除 HDFS 上的文件:

```
[hadoop@server201 ~]$ hdfs dfs -rm  /test/a.txt
```

删除 HDFS 上的目录:

```
[hadoop@server201 ~]$ hdfs dfs -rm -r /test
```

将本地文件上传到 HDFS 上:

```
[hadoop@server201 ~]$ hdfs dfs -copyFromLocal a.txt /test/a.txt
```

使用 put 命令,同样可以将本地文件上传到 HDFS 上:

```
[hadoop@server201 ~]$ hdfs dfs -put a.txt /test/b.txt
```

使用 moveFromLocal 选项,可以同时将本地文件删除:

```
[hadoop@server201 ~]$ hdfs dfs -moveFromLocal a.txt /test/c.txt
```

使用 get/copyToLocal/moveToLocal 选项,可以下载文件到本地:

```
[hadoop@server201 ~]$ hdfs dfs -get /test/c.txt a.txt
[hadoop@server201 ~]$ hdfs dfs -copyToLocal /test/a.txt a1.txt
[hadoop@server201 ~]$ hdfs dfs -moveToLocal /test/a.txt a2.txt
```

2.5 Java 项目访问 HDFS

我们不仅可以使用 hdfs 命令操作 HDFS 文件系统上的文件，还可以使用 Java 代码访问 HDFS 文件系统中的文件。

在 Java 代码中，操作 HDFS 主要通过以下几个主要的类：

- Configuration：用于配置 HDFS。
- FileSystem：表示 HDFS 文件系统。
- Path：表示目录或是文件路径。

可以使用 Eclipse 或使用 IDEA 创建 Java 项目来操作 HDFS 文件系统。其中 Eclipse 是免费软件，IDEA 有两个版本 IC 和 IU 两个版本，其中 IU 为 IDEA Ultimate 为完全功能版本，此版本需要付费后才能使用，不过我们可以选择 IC 版本，C 为 Community 即社区版本的意思，IC 版本为免费版本。后面我们集成开发环境将选择此 IDEA Community 版本。

IDEA 的下载地址为 https://www.jetbrains.com/idea/download/#section=windows。

选择下载 IDEA Community 版本，如图 2-4 所示。

图 2-4

选择下载 zip 版本即可，下载完成后，解压到任意目录下（建议使用没有中文没有空格的目录）。运行 IdeaIC/bin 目录下的 idea64.exe 即可启动 IDEA。接下来，我们就开始创建 Java 项目，并通过 Java 代码访问 HDFS 文件系统。

步骤 01 创建 maven 项目。

打开 IDEA 并选择创建新的项目，如图 2-5 所示。

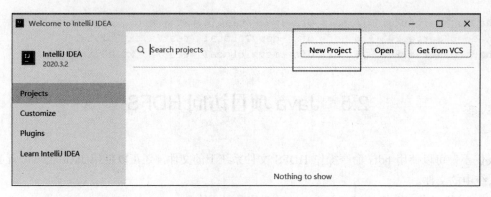

图 2-5

选择创建 Maven 项目，如图 2-6 所示。

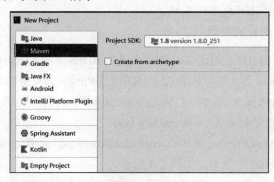

图 2-6

选择项目创建的目录，并输入项目的名称，如图 2-7 所示。

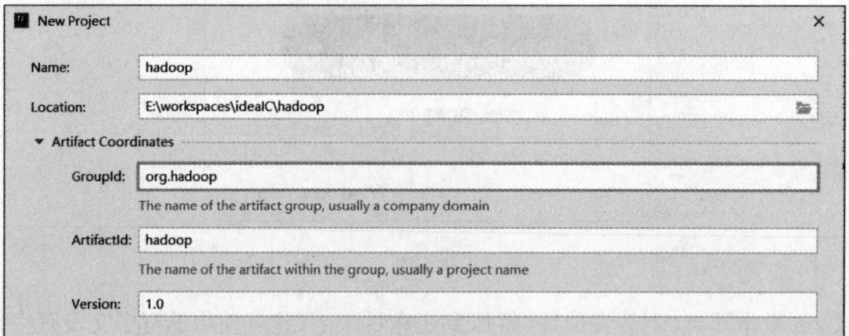

图 2-7

为了方便管理，我们以模块方式来开发，每一个章节可以为一个模块，而 Hadoop 是这些模块的父项目。

所以，在创建完成 Hadoop 项目后，修改 Hadoop 的项目类型为 pom。以下 Hadoop 父项目的 pom.xml 文件部分内容，父项目的<package>类型为 pom。

【代码 2-1】hadoop/pom.xml 文件

```
<groupId>org.hadoop</groupId>
```

```
<artifactId>hadoop</artifactId>
<version>1.0</version>
<packaging>pom</packaging>
```

在父项目中的 dependencyManagement 添加所需要的依赖后，子模块只需要添加依赖名称，不再需要导入依赖的版本。这样父项目就起到了统一管理版本的功能。

```
<dependencyManagement>
    <dependencies>
        <dependency>
            <groupId>org.apache.hadoop</groupId>
            <artifactId>hadoop-client</artifactId>
            <version>3.2.2</version>
        </dependency>
        <dependency>
            <groupId>junit</groupId>
            <artifactId>junit</artifactId>
            <version>4.13.2</version>
        </dependency>
    </dependencies>
</dependencyManagement>
```

现在我们再创建第一个模块。选择 Hadoop 项目，选择创建模块，如图 2-8 所示。

图 2-8

输入模块的名称，如图 2-9 所示。

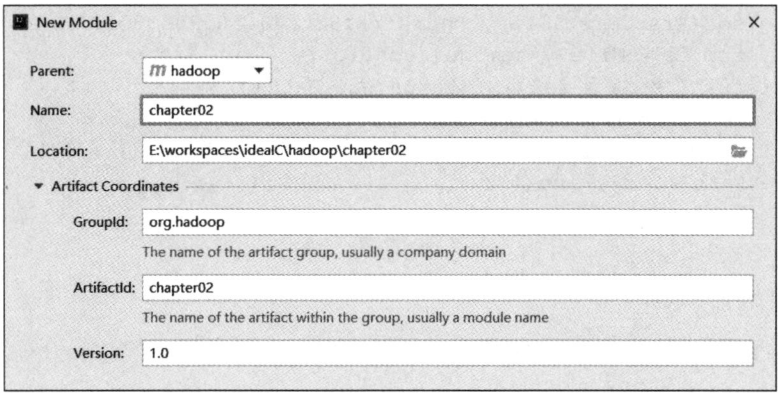

图 2-9

在创建的子模块 chapter01 中，修改 pom.xml 文件，添加以下依赖。注意，只输入 groupId 和 artifactId，不需要输入版本，因为 Hadoop 项目作为父项目管理依赖的版本。

```
<dependencies>
    <dependency>
        <groupId>org.apache.hadoop</groupId>
```

```xml
            <artifactId>hadoop-client</artifactId>
        </dependency>
        <dependency>
            <groupId>junit</groupId>
            <artifactId>junit</artifactId>
            <scope>test</scope>
        </dependency>
</dependencies>
```

查看 hadoop client Aggretator 的依赖关系,如图 2-10 所示。

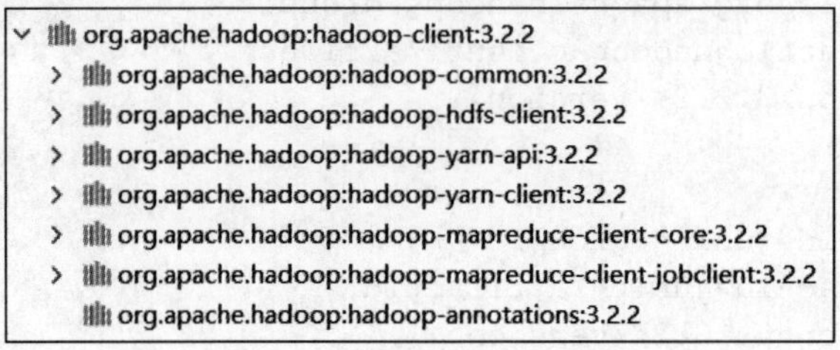

```
✓ ⊞ org.apache.hadoop:hadoop-client:3.2.2
    > ⊞ org.apache.hadoop:hadoop-common:3.2.2
    > ⊞ org.apache.hadoop:hadoop-hdfs-client:3.2.2
    > ⊞ org.apache.hadoop:hadoop-yarn-api:3.2.2
    > ⊞ org.apache.hadoop:hadoop-yarn-client:3.2.2
    > ⊞ org.apache.hadoop:hadoop-mapreduce-client-core:3.2.2
    > ⊞ org.apache.hadoop:hadoop-mapreduce-client-jobclient:3.2.2
      ⊞ org.apache.hadoop:hadoop-annotations:3.2.2
```

图 2-10

至此我们已经可以开发 Java 代码,访问 HDFS 文件系统了。

步骤 02 HDFS 操作示例。

(1) 显示 HDFS 指定目录下的所有目录

显示所有目录,使用 fileSystem.listStatus 方法。

【代码 2-2】Demo01AccessHDFS.java

```java
1.  System.setProperty("HADOOP_USER_NAME", "hadoop");
2.  Configuration config = new Configuration();
3.  config.set("fs.defaultFS", "hdfs://192.168.56.201:8020");
4.  FileSystem fs = FileSystem.get(config);
5.  FileStatus[] stas = fs.listStatus(new Path("/"));
6.  for (FileStatus f : stas) {
7.      System.out.println(f.getPermission().toString() + " "
8.  + f.getPath().toString());
9.  }
10. fs.close();
```

输出的结果如下所示:

```
rwxr-xr-x hdfs://192.168.56.201:8020/test
```

代码说明:

- 第 1 行代码用于设置访问 Hadoop 的用户名。
- 第 2 行代码用于声明一个新的访问配置对象。
- 第 3 行代码设置访问的具体地址。
- 第 4 行代码创建一个文件系统对象。
- 第 5 行~8 行代码为输出根目录下的所有文件或目录,不包含子目录。

- 第 9 行代码关闭文件系统。

（2）显示所有文件

显示所有文件，使用 fileSystem.listFiles 函数，第二个参数 boolean 用于指定是否递归显示所有文件。

【代码 2-3】Demo02ListFiles.java

```
1.  System.setProperty("HADOOP_USER_NAME", "hadoop");
2.  Configuration config = new Configuration();
3.  config.set("fs.defaultFS", "hdfs://192.168.56.201:8020");
4.  FileSystem fs = FileSystem.get(config);
5.  RemoteIterator<LocatedFileStatus> files =
6.  fs.listFiles(new Path("/"), true);
7.  while(files.hasNext()){
8.      LocatedFileStatus file = files.next();
9.      System.out.println(file.getPermission()+" "+file.getPath());
10. }
11. fs.close();
```

添加了 true 参数以后，执行的结果如下所示：

rw-r--r-- hdfs://192.168.56.201:8020/test/a.txt

（3）读取 HDFS 文件的内容

读取 HDFS 上的内容，可以和 fileSystem.open(...) 打开一个文件输入流，然后读取文件流中的内容即可。

【代码 2-4】Demo03ReadFile.java

```
1.  String server = "hdfs://192.168.56.201:8020";
2.  System.setProperty("HADOOP_USER_NAME", "hadoop");
3.  Configuration config = new Configuration();
4.  config.set("fs.defaultFS", server);
5.  try (FileSystem fs = FileSystem.get(config)) {
6.      DataInputStream in = fs.open(new Path(server+"/test/a.txt"));
7.      int len = 0;
8.      byte[] bs = new byte[1024];
9.      while((len=in.read(bs))!=-1){
10.         String str = new String(bs,0,len);
11.         System.out.print(str);
12. }}
```

（4）向 HDFS 写入数据

向 HDFS 写入数据，可以使用 fileSysten.create/append 方法，获取一个 OutputStream，然后向里面输入数据即可。

【代码 2-5】4 Demo04WriteFile.java

```
1.  String server = "hdfs://192.168.56.201:8020";
2.  System.setProperty("HADOOP_USER_NAME", "hadoop");
3.  Configuration config = new Configuration();
```

```
4.    config.set("fs.defaultFS", server);
5.    try (FileSystem fs = FileSystem.get(config)) {
6.        OutputStream out = fs.create(new Path(server+"/test/b.txt"));
7.        out.write("Hello Hadoop\n".getBytes());
8.        out.write("中文写入测试\n".getBytes());
9.        out.close();
10.   }
```

代码输入完成以后，通过 cat 查看文件中的内容：

```
[hadoop@server201 ~]$ hdfs dfs -cat /test/b.txt
Hello Hadoop
中文写入测试
```

其他更多方法不再赘述，这些方法简述如下：

- mkdirs：创建目录。
- create：创建文件。
- checkPath：创建一个文件检查点。
- delete：删除文件。

2.6 winutils

Hadoop 通常运行在 Linux 上，而开发程序通常是 Windows，执行代码时，为了看到更多的日志信息，需要添加 log4j.properties 或 log4j2 的 log4j2.xml。

通过查看 hadoop-client-3.2.2 依赖可知，系统中已经包含了 log4j 1.2 的日志组件，如图 2-11 所示。

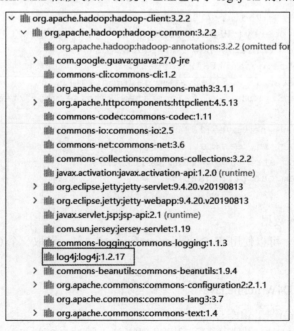

图 2-11

此时只需要添加一个 log4j 与 slf4j 整合的依赖即可，所以添加以下依赖：

```xml
<dependency>
    <groupId>org.slf4j</groupId>
    <artifactId>slf4j-log4j12</artifactId>
    <version>1.7.26</version>
</dependency>
```

同时添加日志文件。直接在 classpath 下创建 log4j.properties 文件，即在项目的 main/resources 目录下添加 log4j.properties 文件即可：

```
log4j.rootLogger = debug,stdout
log4j.appender.stdout = org.apache.log4j.ConsoleAppender
log4j.appender.stdout.Target = System.out
log4j.appender.stdout.layout = org.apache.log4j.PatternLayout
log4j.appender.stdout.layout.ConversionPattern = [%-5p] %d{yyyy-MM-dd HH:mm:ss,SSS} method:%l%n%m%n
```

再次运行上面访问 HDFS 的代码，将出现以下问题：

```
java.io.FileNotFoundException: HADOOP_HOME and hadoop.home.dir are unset
```

意思为 HADOOP_HOME 和 hadoop.home.dir 没有设置。虽然出现上述问题，但程序仍然可以执行成功。出现上述问题的原因是本地没有 Hadoop 的程序。

此时，我们需要在本地，即 Windows 上解压一个相同的 Hadoop 版本，并配置 HADOOP_HOME 环境变量，如笔者将 Hadoop 3.2.2 也同时解压到 Windows 系统的 D:/program 目录下，则配置环境变量为：

```
HADOOP_HOME=D:\program\hadoop-3.2.2
hadoop.home.dir=D:\program\hadoop-3.2.2
```

如果不想配置环境变量，也可以在上述的 Java 代码中，添加上述变量到 System 中，如下所示：

```
System.setProperty("HADOOP_HOME", "D:/program/hadoop-3.2.2");
System.setProperty("hadoop.home.dir", "D:/program/hadoop-3.2.2");
```

同时还需要添加 winutils，可以从 GitHub 或 Gitee 网站上找到相关版本的 winutils。将下载的文件放到 D:\program\hadoop-3.2.2\bin 中即可。再次运行，就没有任何警告信息了。从本书的附带资源中也可以找到相关资源，其位置为 chapter02/软件/wintuils。

2.7　快速 MapReduce 程序示例

MapReduce 为分布式计算模型，分布式计算最早由 Google 提出。MapReduce 将运算的过程分为两个阶段，map 和 reduce 阶段。用户只需要实现 map 和 reduce 两个函数即可。此处先为大家演示一个运行在本地的 MapReduce 程序，后续章节将会重点讲解 MapReduce 的开发。

前面曾经讲过，MapReduce 可以直接在本地模式下运行。在项目中添加 hadoop-mini-cluster 依赖，即可以直接在本地 IDE 环境中运行 MapReduce 程序，而不需要依赖于 Hadoop 集群环境，这个特点在测试开发中非常有用。为了帮助读者开发，我们在代码中加入了开发的步骤，具体代码将会在第 4 章中详细讲解。

在本地运行 Hadoop，需要将 hadoop.dll 文件放到 Windows/system32 目录下，此文件可以在 winutils 目录下找到。

【代码 2-6】Demo02MapReduce.java

```
1.   package org.hadoop;
2.   /**
3.    * MapReduce 示例程序
4.    */
5.   //1：开发一个类，继承 Configured 实现接口 Tool
6.   public class Demo05MapReduce extends Configured implements Tool {
7.   //3：添加 main 函数
8.   public static void main(String[] args) throws Exception {
9.       //7：开始任务
10.      System.setProperty("HADOOP_HOME", "D:/program/hadoop-3.2.2");
11.      System.setProperty("hadoop.home.dir", "D:/program/hadoop-3.2.2");
12.      int res = ToolRunner.run(new Demo05MapReduce(),args);
13.      System.exit(res);
14.  }
15.  //2：实现 run 函数
16.  @Override
17.  public int run(String[] strings) throws Exception {
18.      //6：开发 run 函数内部代码
19.      Configuration conf =getConf();
20.      Job job = Job.getInstance(conf,"WordCount");
21.      job.setJarByClass(getClass());
22.      FileSystem fs = FileSystem.get(conf);
23.      //声明输出目录
24.      Path dest = new Path("D:/a/out");
25.      //如果输出目录已存在则删除
26.      if(fs.exists(dest)){
27.          fs.delete(dest,true);
28.      }
29.      //设置 Mapper 及 Mapper 输出的类型
30.      job.setMapperClass(MyMapper.class);
31.      job.setMapOutputKeyClass(Text.class);
32.      job.setMapOutputValueClass(LongWritable.class);
33.      //设置 Reduce 及 Reduce 的输出类型
34.      job.setReducerClass(MyReduce.class);
35.      job.setOutputKeyClass(Text.class);
36.      job.setOutputValueClass(LongWritable.class);
37.      //设置输入和输出类型
38.      job.setInputFormatClass(TextInputFormat.class);
39.      job.setOutputFormatClass(TextOutputFormat.class);
40.      //设置输入/输出目录
41.      FileInputFormat.addInputPath(job,new Path("D:/a/a.txt"));
42.      FileOutputFormat.setOutputPath(job,dest);
43.      //开始执行任务
```

```
44.     boolean boo = job.waitForCompletion(true);
45.     return boo?0:1;
46. }
47. //4：开发 Mapper 类的实现类
48. public static class MyMapper extends Mapper<LongWritable,Text,Text,LongWritable>{
49.     @Override
50.     protected void map(LongWritable key, Text value, Context context) throws IOException, InterruptedException {
51.         Text outKey = new Text();
52.         LongWritable outValue = new LongWritable(1L);
53.         if(value.getLength()>0){
54.             String[] strs = value.toString().split("\\s+");
55.             for(String str:strs){
56.                 outKey.set(str);
57.                 context.write(outKey,outValue);
58.             }
59.         }
60.     }
61. }
62. //5：开发 Reduce 程序
63. public static class MyReduce extends Reducer<Text,LongWritable,Text,LongWritable>{
64.     @Override
65.     protected void reduce(Text key, Iterable<LongWritable> values, Context context) throws IOException, InterruptedException {
66.         long sum = 0;
67.         LongWritable resultValue = new LongWritable(0);
68.         for(LongWritable v:values){
69.             sum+=v.get();
70.         }
71.         resultValue.set(sum);
72.         context.write(key,resultValue);
73.     }
74. }
75. }
```

运行后，查看 D:\a\out 目录下输出的文件：

```
._SUCCESS.crc
.part-r-00000.crc
_SUCCESS
part-r-00000
```

打开 part-r-0000 即为字符统计的结果，根据源文件不同，统计的结果也会不同，以下仅为参考。默认会以 key 排序，所以输出的数据是以字母作为顺序排列输出的。

```
Configuration  1
Configured     1
Context        2
Demo05MapReduce 1
```

```
Demo05MapReduce(),args);    1
Exception    2
```

2.8 本章小结

本章主要讲解了以下内容：

- Hadoop 单机即伪分布式的环境。
- Hadoop 的配置文件。
- 使用 Hadoop 的脚本 start-dfs.sh 和 start-yarn.sh 分别启动 HDFS 和 YARN。
- Hadoop 启动以后的 5 个进程。
- Hadoop 启动以后，通过 9870、8088 两个端口访问 HDFS 和 MapReduce 的 Web 界面。
- HDFS 的命令行操作。
- Java 操作 HDFS。
- 第一个运行在本地的 MapReduce 程序示例。

第 3 章

HDFS 分布式文件系统

本章主要内容：

- HDFS 的体系结构。
- HDFS 命令。
- RPC 远程调用。
- Hadoop 各进程的功能。

3.1 HDFS 的体系结构

本节将详细详解 HDFS 的组件及其之间的关系，部分图片来自于 Hadoop 的官方网站。图 3-1 展示了客户端如何与 HDFS 进行交互并存取数据的过程。

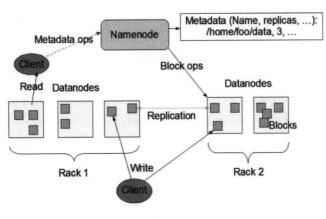

图 3-1

图片引用自 Hadoop 的官方网站：https://hadoop.apache.org/docs/stable/hadoop-project-dist/hadoop-hdfs/HdfsDesign.html。

上图的具体含义如下：

（1）客户端与 NameNode 进行交互，NameNode 通知 Client 将数据存储在哪一台 DataNode 上。由 Client 保存数据到其中一个 DataNode 上，然后 DataNode 会根据元数据信息，再将自己的数据备份到其他 DataNode 上。这样，Client 只要上传到其中一个 DataNode 上，DataNode 之间负责数据副本的复制。NameNode 中 metadata 用于在内存保存数据的元信息。在${hadoop.tmp.dir}/dfs/name 目录下，保存了 edits 和 fsimages 文件，它们分别是日志文件和 metadata 的内存镜像文件。

（2）保存成功以后，由 NameNode 来保存数据的元信息，即什么数据保存到哪几台 DataNode 节点上。

（3）每一个文件块的大小在 Hadoop 2.0 以后为 128MB。可以在 hdfs-site.xml 中配置 dfs.blocksize，其默认值为 134217728 字节，即 128MB。文件块可以理解为将文件分割和存储的大小，如一个文件为 138MB 大小，此文件就会分为两个文件块即 128MB 一块和 10MB 一块，每一块都可以保存到不同的 DataNode 上。在获取文件时多个文件块，会再合并成一个完整的文件。同时，按 HDFS 每一个文件块存在 3 个副本的原则，则 138MB 的文件保存到 HDFS 文件系统上后，大小为 128MB×3 份+10MB×3 份。而在 NameNode 中则保存了这六个文件块的具体位置信息，即元数据信息。

（4）SecondaryNameNode 用于实时地管理 NameNode 中的元数据信息。执行更新和合并的工作。

3.2 NameNode 的工作

在伪分布式的环境中，NameNode 只有一个。但在分布式的环境中，NameNode 被抽象为 NameService。而每一个 NameService 下最多只能有两个 NameNode（Hadoop 3 以后可以有多个）。这种情况下，一个 NameNode 为 Active（活动），另一个为 Standby（备份）。NameNode 主要用于保存元数据、接收客户端的请求。NameNode 的具体工作为：

（1）保存 metadate 元数据，始终在内存中保存 metadata 元数据信息用于处理读请求。由于每一个文件块都会保存一个元数据信息，每个文件的 NameNode 元数据信息大概是 250Byte，4000 万个小文件的元数据信息就需要 4000 万*250Byte≈10GB，也就是说 NameNode 需要 10GB 内存来存储和管理这些小文件的信息。如果存储一个 128MB 的文件，则只需要一个元数据信息，此时 NameNode 的内存，只占 250Byte。如果存储 128 个小文件，每一个文件为 1MB，则此时会有 128 个元数据信息，虽然存储量相同，但 NameNode 管理这些文件块所用的内存并不相同，后者将使用 128×250Byte 的内存大小来保存这些小文件。所以 HDFS 文件系统擅长处理大文件的场景，不擅长处理小文件。

（2）维护 fsimage 文件，也就是 metadate 元数据信息的镜像文件。此文件保存在$hadoop.tmp.dir 所配置的目录（name 目录）下。以下是 fsimage 文件所存在目录及目录下的文件。

```
[hadoop@server201 current]$ pwd
/app/datas/hadoop/dfs/name/current
[hadoop@server201 current]$ ll
总用量 3116
-rw-rw-r-- 1 hadoop hadoop       42 3月   9 22:40 edits_0000000000000000001-000
0000000000000002
```

```
        -rw-rw-r-- 1 hadoop hadoop 1048576 3月  11 22:06 edits_inprogress_00000000000
00000027
        -rw-rw-r-- 1 hadoop hadoop     630 3月  11 22:06 fsimage_0000000000000000026
        -rw-rw-r-- 1 hadoop hadoop      62 3月  11 22:06 fsimage_0000000000000000026.m
d5
        -rw-rw-r-- 1 hadoop hadoop       3 3月  11 22:06 seen_txid
        -rw-rw-r-- 1 hadoop hadoop     218 3月  11 20:06 VERSION
```

（3）当执行写操作时，首先写 editlog 即向 edits 写日志。成功返回以后再写入内存。

（4）SecondaryNameNode 同时维护 fsimage 和 edits 文件以更新 NameNode 的 metadata 元数据。

图 3-2 展示了 NameNode 是如何把元数据保存到磁盘上的。这里有两个不同的文件：

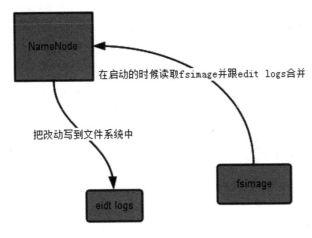

图 3-2

- fsimage 文件：它是在 NameNode 启动时对整个文件系统的快照，是 metadata 的镜像。
- edit logs 文件：每当写操作发生时，NameNode 会首先修改这个文件，然后再去修改 metadata。

元数据信息包含：

- fsimage：为元数据的镜像文件，用于保存一段时间 NameNode 中元数据的信息。
- edits：保存了数据的操作日志。
- fstime：保存最近一次 checkpoint 的时间。fstime 保存在内存中。

这些文件都保存在 dfs.namenode.edits.dir 配置的目录下，如果没有配置此项，则默认保存到 file://${hadoop.tmp.dir}/dfs/name 目录下，正如上面的伪分布式中我们配置的 hadoop.tmp.dir=/app/datas/hadoop，则这些文件都保存到/app/datas/hadoop/dfs/name/current 目录下。

NameNode 始终在内存中保存 metadate 元数据信息。在处理读写数据操作时，会先写 edits 到磁盘，成功返回以后再修改内存中的 metadata。

NameNode 会维护一个 fsimage 文件，此文件是 metadata 保存在磁盘上的镜像文件（Hadoop2/3 中 fsimage 与 metadata 操持实时同步。Hadoop1 不是实时同步）。每隔一段时间 SecondaryNameNode 会合并 fsimage 和 edits 来更新内存中的 matedata。

1. 查看镜像文件

查看 fsimage 文件内容可以使用 hdfs oiv 命令，此命令用于将 fsimage 文件转成可读的 XML 文件。oiv 的具体含义，可以通过 help 命令获取：

```
[hadoop@server201 current]$ hdfs oiv -h
Usage: bin/hdfs oiv [OPTIONS] -i INPUTFILE -o OUTPUTFILE
Offline Image Viewer
View a Hadoop fsimage INPUTFILE using the specified PROCESSOR,
saving the results in OUTPUTFILE.
```

通过上述的命令可以看到 oiv 的含义为 Offline Image View，即查看离线文件。oiv 常用参数为：

```
Required command line arguments:
-i,--inputFile <arg>   FSImage or XML file to process.

Optional command line arguments:
-o,--outputFile <arg>  Name of output file. If the specified
                       file exists, it will be overwritten.
                       (output to stdout by default)
                       If the input file was an XML file, we
                       will also create an <outputFile>.md5 file.
-p,--processor <arg>   Select which type of processor to apply
                       against image file. (XML|FileDistribution|
                       ReverseXML|Web|Delimited)
                       The default is Web.
```

参数说明：

- -i,--inputFile：用于指定 FsImage 或 XML 文件名称。
- -o,--outputFile：输出文件名称，如果文件已经存在则会覆盖，同时会生成一个 *.md5 比对文件。
- -p,--processor：处理方式，默认为 Web，建议使用 XML。

现在我们使用 oiv 命令查看一个已经存在的 fsimage 文件，其命令如下：

```
[hadoop@current]$ hdfs oiv -i fsimage_0000000000000000024 -o ~/fsimage.xml -p XML
```

上面的命令用于将 fsimage 文件转成 XML 文件，转换成功后，就可以使用 vim 查看 fsimage 文件的内容了。打开此文件，并格式化，其内容如下（根据不同的环境内容会有不同，仅作参考）：

```xml
<?xml version="1.0" encoding="utf-8"?>
<fsimage>
    <version>
        <layoutVersion>-65</layoutVersion>
        <onDiskVersion>1</onDiskVersion>
        <oivRevision>7a3bc90b05f257c8ace2f76d74264906f0f7a932</oivRevision>
    </version>
    ...
</fsimage>
```

可见在 fsimage 文件中保存了版本、文件、节点等信息。

2. 查看日志文件

hdfs oev 命令用于查看 edis 日志文件，同样地，可以通过 help 命令，查看它的具体使用方法：

```
[hadoop@server201 current]$ hdfs oev -h
Usage: bin/hdfs oev [OPTIONS] -i INPUT_FILE -o OUTPUT_FILE
Offline edits viewer
Parse a Hadoop edits log file INPUT_FILE and save results
in OUTPUT_FILE.
```

可见 oev 的含义为 Offline Edits View，即查看离线 edits 文件。现在我们使用此命令将 edits 文件转成可读的 XML 文件：

```
$ hdfs oev -i edits_0000000000000000013-0000000000000000014 -o ~/edits.xml -p XML
```

查看得到的 XML 文件内容如下，可见此文件中只记录了操作的事务日志信息。

```xml
<?xml version="1.0" encoding="UTF-8" standalone="yes"?>
<EDITS>
    <EDITS_VERSION>-65</EDITS_VERSION>
    <RECORD>
        <OPCODE>OP_START_LOG_SEGMENT</OPCODE>
        <DATA>
            <TXID>13</TXID>
        </DATA>
    </RECORD>
    <RECORD>
        <OPCODE>OP_END_LOG_SEGMENT</OPCODE>
        <DATA>
            <TXID>14</TXID>
        </DATA>
    </RECORD>
</EDITS>
```

3. 日志文件和镜像文件的操作过程

在 edits 文件中，有一个名为 edits_inprogress_开头的文件，表示是正在操作的日志文件。如当前目录下，我们正有一个这样的日志文件，它的名称为：

```
edits_inprogress_0000000000000000028
```

此文件将实时记录我们对 HDFS 文件的操作，如我们现在删除一个文件：

```
$ hdfs dfs -rm /test/a.txt
```

然后通过 oev 命令，将 edits_inprocess_开始的文件转成 XML 文件，并读取里面的内容。

```
$ hdfs oev -i edits_inprogress_0000000000000000028 -o ~/inprocess.xml -p XML
```

我们就会在文件的最后，发现记录删除文件的操作记录：

```
<RECORD>
    <OPCODE>OP_DELETE</OPCODE>
    <DATA>
        <TXID>30</TXID>
        <LENGTH>0</LENGTH>
        <PATH>/test/a.txt</PATH>
        <TIMESTAMP>1615559516088</TIMESTAMP>
        <RPC_CLIENTID>68c4f333-8bad-472a-8990-4551ce0e5c2c</RPC_CLIENTID>
        <RPC_CALLID>3</RPC_CALLID>
    </DATA>
</RECORD>
```

其中 OPCODE 的值为 OP_DELETE，即为删除文件。在 DATA 节点中则记录了文件的详细信息。可见 edits 文件，会实时记录文件的操作日志。

4. 合并 fsimage 镜像文件

默认情况下，关闭打开 HDFS 进程，会自动合并保存 fsimage 文件到磁盘。不过，也可以通过以下命令，将内存中的 matedata 保存为新的 fsimage 文件，首先查看目前的日志文件，可知最新的版本为 28，即 edits_inprocess_xxx28 这个数字，就是最新的日志版本。而 fsimage 即内存中的 matedata 保存到磁盘上镜像文件总是比值小一个版本为 27，如图 3-3 所示。

```
      42 3月      9 22:40 edits_0000000000000000001-0000000000000000002
 1048576 3月      9 22:40 edits_0000000000000000003-0000000000000000003
     611 3月     10 21:35 edits_0000000000000000004-0000000000000000012
      42 3月     10 22:35 edits_0000000000000000013-0000000000000000014
 1048576 3月     10 22:42 edits_0000000000000000015-0000000000000000021
      88 3月     11 21:06 edits_0000000000000000022-0000000000000000024
      42 3月     11 22:06 edits_0000000000000000025-0000000000000000026
 1048576 3月     11 22:06 edits_0000000000000000027-0000000000000000027
 1048576 3月     12 22:31 edits_inprogress_0000000000000000028
     630 3月     11 22:06 fsimage_0000000000000000026
      62 3月     11 22:06 fsimage_0000000000000000026.md5
     630 3月     12 22:27 fsimage_0000000000000000027
      62 3月     12 22:27 fsimage_0000000000000000027.md5
       3 3月     12 22:27 seen_txid
     218 3月     12 22:27 VERSION
```

图 3-3

或通过输出 seen_txid 可知，如果下一次合并则版本号为 32。

```
[hadoop@server201 current]$ cat seen_txid
32
```

上面说明 fsimage_0000000000000000027 还没有与 edits_0000000000000000028 合并。现在我们使用以下命令，执行合并工作。

先进入安全模式，通过 dfsadmin 可以将 HDFS 进入安全模式，进入安全模式后，将不能操作 HDFS 文件系统中的文件，一般在 HDFS 刚刚启动时，会有一段时间为安全模式，等所有的 fsimage 文件加载到内存后，将会自动退出安全模式。

使用 dfsadmin 命令让 HDFS 进入安全模式：

```
[hadoop@server201 current]$ hdfs dfsadmin -safemode enter
Safe mode is ON
```

也可以通过 dfsadmin get 查看当前是否处于安全模式下：

```
[hadoop@server201 current]$ hdfs dfsadmin -safemode get
Safe mode is ON
```

保存 matedata 数据到 fsimage 中：

```
[hadoop@server201 current]$ hdfs dfsadmin -saveNamespace
Save namespace successful
```

退出安全模式：

```
[hadoop@server201 current]$ hdfs dfsadmin -safemode leave
Safe mode is OFF
```

现在再来查看日志文件，已经发生变化，如图 3-4 所示。说明已经将内存中的元数据保存到新的 fsimage 镜像文件中。

```
 9 22:40 edits_0000000000000000001-0000000000000000002
 9 22:40 edits_0000000000000000003-0000000000000000003
10 21:35 edits_0000000000000000004-0000000000000000012
10 22:35 edits_0000000000000000013-0000000000000000014
10 22:42 edits_0000000000000000015-0000000000000000021
11 21:06 edits_0000000000000000022-0000000000000000024
11 22:06 edits_0000000000000000025-0000000000000000026
11 22:06 edits_0000000000000000027-0000000000000000027
12 22:47 edits_0000000000000000028-0000000000000000031
12 23:08 edits_0000000000000000032-0000000000000000033
12 23:08 edits_inprogress_0000000000000000034
12 22:27 fsimage_0000000000000000027
12 22:27 fsimage_0000000000000000027.md5
12 23:08 fsimage_0000000000000000033
12 23:08 fsimage_0000000000000000033.md5
12 23:08 seen_txid
12 23:08 VERSION
```

图 3-4

3.3　SecondaryNameNode

SecondaryNameNode 为 HA（高可用）的一个解决方案。在伪分布式中一般只有一个 NameNode，一旦 NameNode 宕机，内存中的元数据信息也将随之丢失，而 NameNode 中的元数据信息就像是图书中的目录一样，记录了所有文件存储的信息。如果再次启动 NameNode 不能找回元信息，将会无法找到之前存储的文件。而此时 SecondaryNameNode 就扮演了非常重要的角色，它实时将 NameNode 的元数据保存到磁盘上，NameNode 重新启动时，会读取磁盘上的元数据信息到内存，从而将损失降到最小。SecondaryNameNode 的职责是合并 NameNode 的 edit logs 到 fsimage 文件中，如图 3-5 所示。在正式的集群中没有 SecondaryNameNode 进程，而是多个 NameNode 进程互为主备。

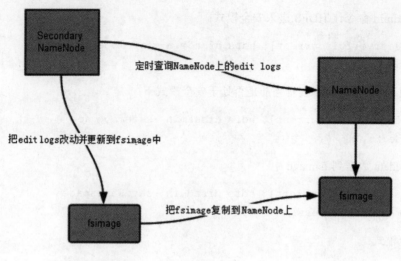

图 3-5

图 3-5 展示了 SecondaryNameNode 是如何工作的:

(1) 它定时到 NameNode 去获取 edit logs (内存中的 editlog 即为 metadata),并更新到 fsimage 中。具体执行合并更新的时间可以通过以下两个参数在 hdfs-site.xml 文件中进行配置,以最先达到的值为执行的时间点。

- 时间参数:dfs.namenode.checkpoint.period 默认值为 3600s,即 3600 秒。
- 事务参数:dfs.namenode.checkpoint.txns 默认值为 1000000,即 100 万事务后将执行合并。

(2) NameNode 在下次重启时会使用这个新的 fsimage 文件和 edists 文件进行合并,将合并到内存中的 fsimages 的元数据信息重新保存到磁盘上。

(3) SecondaryNameNode 的整个目的是在 HDFS 中提供一个检查点。它只是 NameNode 的一个助手。

现在,我们了解了 SecondaryNameNode 所做的不过是帮助 NameNode 更好的工作。它并不是要取代 NameNode 也不是 NameNode 的备份。

3.4 DataNode

DataNode 作为数据存储的节点,并实时与 NameNode 通过心跳机制保持通信。

DataNode 的功能如下:

(1) 提供真实的存储服务。

(2) 在 Hadoop 2.0 及以上的版本中,每一个文件块的大小为 128MB,文件默认块大小为 128MB,如果一个文件的大小没有 128MB,则上传的文件将会占用一个实际大小的空间。如果文件大于 128MB,则文件将会被分割成多个文件块。读者可以通过上传一个大于 128MB 的文件后查看一下上传后的文件是否分成多个文件的形式保存。文件的块大小可以在 hdfs-site.xml 文件中添加配置 dfs.blocksize，默认值为 134217728，即 128MB。

（3）如果在 core-site.xml 中配置了 hadoop.tmp.dir，会将真实的数据保存到${hadoop.tmp.dir}/data 目录下。如果没有配置，则默认会将数据保存到/tmp/hadoop-${user.name}中。在上面的目录下有一个 data 目录，里面就是保存 HDFS 真实数据的位置。

（4）默认的副本为 3 个，在 hdfs-site.xml 中配置 dfs.replication 可以修改默认副本数量，最大小为 512MB，默认值为 3。

3.5　HDFS 的命令

在 Hadoop 1.x 版本中，使用 hadoop 命令管理 HDFS 文件系统。在 Hadoop 2.x 版本中，使用 hdfs 命令管理 HDFS 文件系统。

以下是 Hadoop 1.x 版本的命令，现在依然可以使用：

```
# hadoop fs -ls /
Found 2 items
drwxr-xr-x   - root supergroup          0 2018-12-09 21:45 /test
drwx------   - root supergroup          0 2018-12-09 20:46 /tmp
```

也可以省去 hdfs://server201:8020，直接输入"/"（斜线）即可：

```
[root@server51 ~]# hdfs dfs -ls /
Found 2 items
drwxr-xr-x   - root supergroup          0 2018-12-09 21:45 /test
drwx------   - root supergroup          0 2018-12-09 20:46 /tmp
```

以下是几个常用的命令。显示服务器文件列表：

```
hdfs  dfs -ls /
```

将本地文件复制到 HDFS 上去：

```
$hdfs dfs -copyFromLocal  ~/home/wangjian/some.txt  /some.txt
```

查看服务器上的文件内容：

```
$hdfs dfs -cat /some.txt
```

从服务器下载文件到本地：

```
$hdfs dfs -copyToLocal /test1.txt test1.txt
```

服务器文件和文件夹计数：

```
$hdfs dfs -count /
```

向服务器上传文件：

```
$hdfs dfs -put test1.txt /test2.txt
```

从服务器获取文件到本地：

```
$hdfs dfs -get /test2.txt test3.txt
```

3.6 RPC 远程过程调用

在 Hadoop 中，多个进程（NameNode/DataNode 等）之间数据的传递和访问都是通过 RPC 实现的。RPC 是不同进程之间方法调用的解决方案。RPC 调用的原理是通过网络代理实现远程方法的调用。这些功能已经被 Hadoop 封装，直接使用 Hadoop 提供的类即可以实现 RPC 远程调用。

由于被调用的类是通过动态代理实现的，所以必须拥有一个接口。且接口上必须拥有一个 public static final long versionID=xxxxL 的声明，即序列化的 versionID 字段。

以下简单通过 Hadoop RPC 代码调用示例，演示 RPC 在多个进程之间调用的过程。

步骤 01 开发一个接口。

在接口中，必须拥有一个 versionID 唯一的标识当前接口。

【代码 3-1】IHello.java

```
1.  package org.hadoop.rpc.service;
2.  public interface IHello {
3.      //必须声明一个 versionID 值是任意的 Long 值
4.      long versionID = 937530L;
5.      //定义一个方法，用于远程测试调用
6.      String say(String name);
7.  }
```

步骤 02 开发实现类，并实现接口中的方法。

实现接口，且实现接口中的方法。声明一个类，并声明一个方法，返回一个任意的字符串。一个类如果希望被远程 RPC 调用，这个类必须实现一个接口，因为内部的原理是 JDK 动态代理。

【代码 3-2】HelloImpl.java

```
1.  package org.hadoop.rpc.service;
2.  import java.time.LocalDateTime;
3.  public class HelloImpl implements IHello{
4.      @Override
5.      public String say(String name) {
6.          String str = "Hello "+name+", 当前时间是："+ LocalDateTime.now();
7.          return str;
8.      }
9.  }
```

步骤 03 开发服务器。

服务器通过 RPC.Builder 来创建服务，并通过一个指定的端口对外暴露需要被调用的类。

【代码 3-3】RpcServer.java

```
1.  package org.hadoop.rpc.server;
2.  public class RpcServer {
3.    public static void main(String[] args) throws Exception {
4.      System.setProperty("hadoop.home.dir","D:/program/hadoop-3.2.2");
5.      RPC.Server server = new RPC.Builder(new Configuration())
6.              .setProtocol(IHello.class)//设置协议即接口
7.              .setInstance(new HelloImpl())//设置实例类
8.              .setBindAddress("127.0.0.1")//设置服务器地址
9.              .setPort(5678)//设置服务端口
10.             .build();
11.     server.start();//启动 rpc 服务
12.   }
13. }
```

开发完成 RpcServer 后，右击运行，即可启动 RpcServer。启动后，将会占用 5678 端口，并等待客户端的连接，输出的信息如下所示，即表示启动成功。

```
Starting Socket Reader #1 for port 5678
IPC Server listener on 5678: starting
```

步骤 04 开发客户端。

使用 RPC.getProxy 获取本地的一个代理，但是接口类必须与服务器的接口类保持一样，如果调用方是另一个项目，请将接口类复制到调用项目中即可。

【代码 3-4】RpcClient.java

```
1.  package org.hadoop.rpc.client;
2.  public class RpcClient {
3.    public static void main(String[] args) throws Exception {
4.      System.setProperty("hadoop.home.dir","D:/program/hadoop-3.2.2");
5.      IHello hello =
6.          RPC.getProxy(IHello.class,//指定协议或接口
7.              1L,//指定客户端 ID，任意 Long 值
8.              new InetSocketAddress("127.0.0.1", 5678),
9.              new Configuration());
10.     String str = hello.say("hadoop");
11.     System.out.println("返回的数据是："+str);
12.   }
13. }
```

最后，请运行上面的程序代码，如果可以正常返回数据，则说明 RPC 远程调用成功。RPC 调用的过程就是通过服务器地址和端口实现远程类方法的调用。

3.7 本章小结

Hadoop 默认配置在 core-default.xml 文件中，此文件包含在 hadoop-common.jar 包中。里面的部

分配置如下:

```
1.   <property>
2.     <name>hadoop.common.configuration.version</name>
3.     <value>3.0.0</value>
4.     <description>version of this configuration file</description>
5.   </property>
6.   <property>
7.     <name>hadoop.tmp.dir</name>
8.     <value>/tmp/hadoop-${user.name}</value>
9.     <description>A base for other temporary directories.</description>
10.  </property>
```

说明:

- 第3行代码说明了 Hadoop 的大版本号。
- 第8行代码如果没有配置 hadoop.tmp.dir, 则默认将数据保存到的目录。

建议不要修改 core-default.xml 文件中的内容, 如果需要可以修改${HADOOP_HOME}/etc/hadoop/core-site.xml 中的内容。

在 hadoop-hdfs.jar 中包含 hdfs-default.xml 文件, 里面保存了默认的 HDFS 配置, 例如:

```
1.   <property>
2.     <name>dfs.replication</name>
3.     <value>3</value>
4.   </property>
5.   <property>
6.     <name>dfs.blocksize</name>
7.     <value>134217728</value>
8.   </property>
```

说明:

- 第3行代码为默认副本数量。
- 第7行代码为默认块大小, 值为128MB。

第 4 章

分布式运算框架 MapReduce

本章主要内容：

- MapReduce 的执行原理、执行过程。
- 数据类型及数据格式。
- Writable 接口与序列化机制。
- WritableComparable 实现排序。
- 默认 Mapper 和默认 Reducer。
- 倒排索引。
- MapReduce 的源码分析。

4.1 MapReduce 的运算过程

MapReduce 为分布式计算模型，分布式计算最早由 Google 提出。MapReduce 将运算的过程分为两个阶段，map 和 reduce 阶段。用户只需要实现 map 和 reduce 两个函数即可。这两个函数参数的形式都是以 Key-Value 的形式成对出现，Key 为输出或输出的信息，Value 为输入或输出的值。

图 4-1 展示了 MapReduce 的运算过程。将大任务交给多个机器分布式进行计算，然后再进行汇总合并。

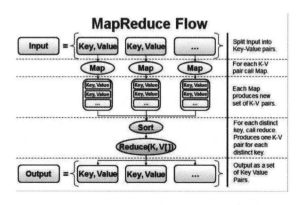

图 4-1

图 4-1 介绍如下：

第 1 行，Input 为输入源，因为输入源是以分布式形式保存到 HDFS 上的，所以可以同时开启多个 Mapper 程序，同时读取数据。读取数据时，Mapper 将接收两个输入的参数，第一个 Key 为读取到的文件的行号，Value 为读取到的这一行的数据。

第 2~3 行，Mapper 在处理完成以后，也将输出 Key-Value 对形式的数据。

第 4 行，是 Reducer 接收 Mapper 输入的数据，在接收数据之前进行排序操作，这个排序操作我们一般称为 shuffle。注意，Reducer 接收到的 Value 值是一个数组，多个重复 Key 的 Value 会在 Reducer 程序中合并成数组。

第 5 行，是 Reducer 的输出，也是以 Key-Value 对象的形式对外输出到文件或其他存储设备中。

以 WordCount 为示例，再为读者讲解一下 MapReduce 的过程，如图 4-2 所示。

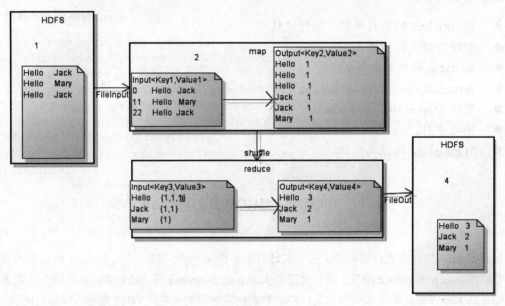

图 4-2

（1）标注为 1 的部分为 Hadoop 的 HDFS 文件系统中的文件，即被处理的数据应该首先保存到 HDFS 文件系统上。

（2）标注为 2 的部分，将接收 FileInputFormat 的输入数据。在处理 WordCount 示例时，接收到的数据 key1 为 LongWritable 类型，即为字节的偏移量。比如，标注为 2 的部分中第一行输入为 0，其中 0 为字节 0 下标的开始；第二行为 11，则 11 为文本中第二行字节的偏移量，以此类推。而 Value1 则为 Text 即文本类型，其中第一行 Hello Jack 为读取的第一行的数据，以此类推。然后，此时我们将开发代码对 Value1 的数据进行处理，以空格或是\t 作为分割，分别将 Hello 和 Jack 依次输出。此时每一次输出的算是一个字符，所以在 Map 中的输出格式 Key2 为 Text 类型，而 Value2 则为 LongWritable 类型。

（3）标注为 3 的部分，接收 Map 的输出，所以 Key3 和 Value3 的类型应该与 Key2 和 Value2 的类型一致。现在我们只需要将 Value 中的值相加，就可以得到 Hello 出现的次数。然后直接输出给 Key4 和 Value4。

如果你已经理解了这个 MapReduce 的过程，接下来就可以快速开发 WordCount 代码了。

注意：LongWritable 和 Text 为 Hadoop 中的序列化类型。可以简单理解为 Java 中的 Long 和 String。

4.2 WordCount 示例

为了使读者快速掌握 MapReduce，本节再次为大家开发和演示 WordCount 示例程序，并以本地运行和服务器运行的方式分别部署，使读者更深入了解 MapReduce 的开发、运行和部署。

前面已经介绍 MapReduce 程序可以运行在本地，也可以打包后运行在 Hadoop 集群上。之前已经开发过运行在本地的 MapReduce 程序，这里我们将使用打包的方式将程序打包后放到 Hadoop 集群上运行。

步骤 01 创建 Java 项目并添加依赖。

创建 Java 项目，并添加以下依赖。注意，本次以添加的依赖为 hadoop-minicluster，且设置 scope 的值为 provided（意思是，在打包时将不会被打包到依赖的 jar 包中）。

```xml
1.  <dependency>
2.      <groupId>org.apache.hadoop</groupId>
3.      <artifactId>hadoop-minicluster</artifactId>
4.      <scope>provided</scope>
5.  </dependency>
```

步骤 02 开发 WordCount 的完整代码。

【代码 4-1】WordCount.java

```java
1.  package org.hadoop;
2.  public class WordCount extends Configured implements Tool {
3.      public static void main(String[] args) throws Exception {
4.          int result = ToolRunner.run(new WordCount(), args);
5.          System.exit(result);
6.      }
7.      private static String server = "hdfs://server201:8020";
8.      public int run(String[] args) throws Exception {
9.          if (args.length != 2) {
10.             System.err.println("usage: " + this.getClass().getSimpleName() + " <inPath> <outPath>");
11.             ToolRunner.printGenericCommandUsage(System.out);
12.             return -1;
13.         }
14.         Configuration config = getConf();
15.         config.set("fs.defaultFS", server);
16.         //指定 resourcemanager 的地址
17.         config.set("yarn.resourcemanager.hostname", "server201");
18.         config.set("dfs.replication", "1");
19.         config.set("dfs.permissions.enabled", "false");
20.         FileSystem fs = FileSystem.get(config);
```

```
21.         Path dest = new Path(server + args[1]);
22.         if (fs.exists(dest)) {
23.             fs.delete(dest, true);
24.         }
25.         Job job = Job.getInstance(config,"WordCount");
26.         job.setJarByClass(getClass());
27.         job.setMapperClass(WordCountMapper.class);
28.         job.setReducerClass(WordCountReducer.class);
29.         job.setOutputKeyClass(Text.class);
30.         job.setOutputValueClass(LongWritable.class);
31.         FileInputFormat.addInputPath(job, new Path(server + args[0]));
32.         FileOutputFormat.setOutputPath(job, dest);
33.         boolean boo = job.waitForCompletion(true);
34.         return boo ? 0 : 1;
35.     }
36.     public static class WordCountMapper extends Mapper<LongWritable, Text, Text, LongWritable> {
37.         private LongWritable count = new LongWritable(1);
38.         private Text text = new Text();
39.         @Override
40.         public void map(LongWritable key, Text value, Context context) throws IOException, InterruptedException {
41.             String str = value.toString();
42.             String[] strs = str.split("\\s+");
43.             for (String s : strs) {
44.                 text.set(s);
45.                 context.write(text, count);
46.             }
47.         }
48.     }
49.     public static class WordCountReducer extends Reducer<Text, LongWritable, Text, LongWritable> {
50.         @Override
51.         public void reduce(Text key, Iterable<LongWritable> values,Context context) throws IOException, InterruptedException {
52.             long sum = 0;
53.             for (LongWritable w : values) {
54.                 sum += w.get();
55.             }
56.             context.write(key, new LongWritable(sum));
57.         }
58.     }
59. }
```

上例代码中,由于我们声明了完整的地址,所以可以在本地运行测试。在本地运行测试需要输入两个参数。选择 IDEA 的 run > Edit Configurations,并在 Program Arguments 位置输入读取文件的地址和输出结果的目录,如图 4-3 所示。

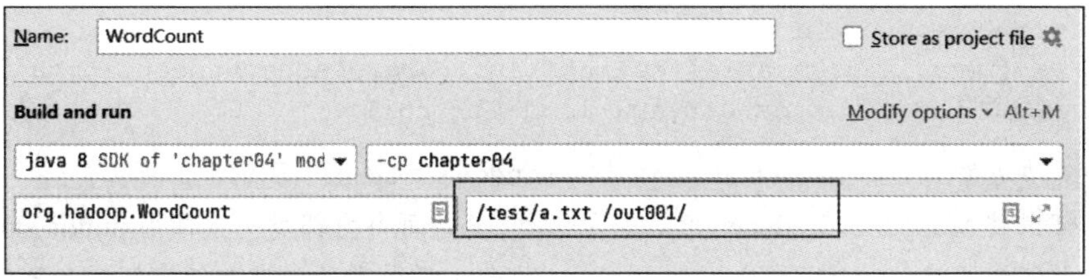

第 4 章　分布式运算框架 MapReduce | 59

图 4-3

在本地环境下直接运行，并查看 HDFS 上的结果目录，WordCount 程序已经将结果输出到指定的目录中。

```
[hadoop@server201 ~]$ hdfs dfs -ls /out001
Found 2 items
-rw-r--r--   1 mrchi supergroup          0 2021-03-13 22:14 /out001/_SUCCESS
-rw-r--r--   1 mrchi supergroup        520 2021-03-13 22:14 /out001/part-r-00000
```

步骤 03 使用 Maven 打包程序。

在 IDEA 右侧栏的 Maven 视图中，单击 package 并运行，可以得到一个 jar 包，如图 4-4 所示。

图 4-4

打完的包可以在 target 目录下找到，将 jar 包上传到 server201 服务器的/root 目录下，并使用 yarn jar 执行。

使用 yarn jar 执行，使用以下命令：

```
$ yarn jar chapter04-1.0.jar org.hadoop.WordCount /test/a.txt /out002
```

查看执行结果，即为单词统计的结果。根据处理的文件不同，这个结果文件的内容会有所不同。

```
[root@server201 ~]# hdfs dfs -cat /out002/* | head
->    4
0     2
1     4
```

至此，你已经学会如何在本地及打包到服务器上运行 MapReduce 程序了。接下来，我们将详解 MapReduce 的更多细节。

注意：在本地运行时，有可能会出现以下错误：

```
Exception in thread "main"
    java.lang.UnsatisfiedLinkError: org.apache.hadoop.io.nativeio.Nati
veIO$Windows.access0(Ljava/lang/String;I)Z
```

解决方案是：将 hadoop.dll 文件放到 windows/system32 目录下即可。

可以将 Mapper 和 Reducer 开发成内部类，但这两个内部类必须使用 public static 修饰符。上例的程序，在 IDEA 中执行 package 打包，将得到一个没有任何依赖，只有 WordCount 代码的 jar 文件。之后，就可以发布到 Linux 上并在 Hadoop 集群中执行。因为在 Linux 上已经存在了 Hadoop 的所有依赖包，所以不需要再将 Hadoop 的所有依赖都打包到 jar 文件中去。

4.3 自定义 Writable

在 Hadoop 中，LongWritable、Text 都是被序列化的类，它们都是 org.apache.hadoop.io.Writable 的子类。也只有被序列化的类，才可以在 Mapper 和 Reducer 之间传递。在实际开发中，为了适应不同业务的需求，有时必须自己开发 Writable 类的子类，以实现 Hadoop 中的个性化开发。以下是 JDK 的序列化与 Hadoop 序列化的比较。

- JDK 的序列化接口——java.io.Serializable：用于将对象转换成字节流输出，即为序列化，再将字节流转换成对象，即为反序列化。
- Hadoop 的序列化接口——org.apache.hadoop.io.Writable：它的特点是紧凑（高效地使用存储空间）、快速（读/写开销小）、可扩展（可以透明的读取数据）、互操作（支持多语言）。

Writable 接口的两个主要方法：write(DataOutput) 将成员变量按顺序写出，readFields(DataInput) 顺序读取成员变量的值。以下是 Writable 的源代码：

```
package org.apache.hadoop.io;
public interface Writable {
    void write(DataOutput out) throws IOException;
    void readFields(DataInput in) throws IOException;
}
```

Writable 接口的基本实现，就是在 write/readFields 中顺序写出和读取成员变量的值。以下代码是一个实现了 Writable 接口的具体类，此类中省略了 setters 和 getters 方法，请注意 write 和 readFields 中的代码，必须按顺序读写数据。

【代码 4-2】WritableDemo.java

```
1.  package org.hadoop;
2.  public class WritableDemo implements Writable {
3.      private String name;
4.      private Integer age;
5.      @Override
6.      public void write(DataOutput out) throws IOException {
```

```
7.         out.writeUTF(name);
8.         out.writeInt(age);
9.     }
10.    @Override
11.    public void readFields(DataInput in) throws IOException {
12.        name = in.readUTF();
13.        age = in.readInt();
14.    }
15. }
```

在自定义的类实现接口 Writable 以后，就可以将这个类作为 Key 或是 Value 放到 Mapper 或是 Reducer 中当作参数。

现在我们来自己定义一个序列化类，用于统计文本文件中每一行字符的数量。根据这个要求，我们需要定义 Mapper 的输出类型为自定义的这个 Writable，而 Mapper 的输出则为 Reduce 的输入。

步骤 01 创建目标读取文件。

首先定义一个文本文件，并输入若干行内容。如创建一个 a.txt 文件并存储在 D:/目录下。

```
Hello This Is First Line .
This Example show how
to implements Writable
```

步骤 02 创建 Writable 类的子类。

由于我们的工作是要读取这一行的数据和计算这一行有多少字符，所以应该定义两个成员变量，如【代码 4-3】所示。注意，在【代码 4-3】中，我们实现的是 Writable 的子类 WritableComparable<T>，因为我们要将此类作为输出的 key 值，则输出的 key 必须实现排序的功能，此类 WritableComparable 的 comparaTo 方法可以实现排序。

【代码 4-3】LineCharCountWritable.java

```
1.  package org.hadoop.writable;
2.  public class LineCharCountWritable implements
3.              WritableComparable<LineCharCountWritable> {
4.      private String line;
5.      private Integer count;
6.      @Override
7.      public void write(DataOutput out) throws IOException {
8.          out.writeUTF(line);
9.          out.writeInt(count);
10.     }
11.     @Override
12.     public void readFields(DataInput in) throws IOException {
13.         line = in.readUTF();
14.         count = in.readInt();
15.     }
16.     @Override
17.     public int compareTo(LineCharCountWritable o) {
18.         return this.count - o.getCount();
19.     }
```

步骤 03 在 MapReduce 中使用自定义 Writable。

在项目中使用自定义的 Writable 子类，参见【代码 4-4】。

【代码 4-4】LineCharCountMR.java

```java
1.   package org.hadoop.writable;
public class LineCharCountMR extends Configured implements Tool {
2.       public static void main(String[] args) throws Exception {
3.           int run = ToolRunner.run(new LineCharCountMR(), args);
4.           System.exit(run);
5.       }
6.       @Override
7.       public int run(String[] args) throws Exception {
8.           if (args.length < 2) {
9.               System.out.println("参数错误，使用方法：LineCharCountMR <Input> <Output>");
10.              ToolRunner.printGenericCommandUsage(System.out);
11.              return 1;
12.          }
13.          Configuration config = getConf();
14.          FileSystem fs = FileSystem.get(config);
15.          Path dest = new Path(args[1]);
16.          if (fs.exists(dest)) {
17.              fs.delete(dest, true);
18.          }
19.          Job job = Job.getInstance(config, "LineChar");
20.          job.setJarByClass(getClass());
21.          job.setMapperClass(LineMapper.class);
22.          job.setMapOutputKeyClass(LineCharCountWritable.class);
23.          job.setMapOutputValueClass(NullWritable.class);
24.          job.setReducerClass(LineReducer.class);
25.          job.setOutputKeyClass(Text.class);
26.          job.setOutputValueClass(NullWritable.class);
27.          FileInputFormat.addInputPath(job, new Path(args[0]));
28.          FileOutputFormat.setOutputPath(job, dest);
29.          boolean b = job.waitForCompletion(true);
30.          return b ? 0 : 1;
31.      }
32.      //注意最后一个参数为 NullWritable，可以理解为 Null
33.      public static class LineMapper extends Mapper<LongWritable, Text, LineCharCountWritable, NullWritable> {
34.          private LineCharCountWritable countWritable = new LineCharCountWritable();
35.          @Override
36.          protected void map(LongWritable key, Text value, Context context) throws IOException, InterruptedException {
37.              String line = value.toString();
38.              if (StringUtils.isBlank(line)) {
```

```
39.            return;
40.        }
41.        Integer charCount = line.length();
42.        countWritable.setLine(line);
43.        countWritable.setCount(charCount);
44.        context.write(countWritable, NullWritable.get());
45.    }
46. }
47. public static class LineReducer extends Reducer<LineCharCountWritable, NullWritable, Text, NullWritable> {
48.    private Text text = new Text();
49.    @Override
50.    protected void reduce(LineCharCountWritable key, Iterable<NullWritable> values, Context context) throws IOException, InterruptedException {
51.        text.set(key.getLine() + "\t" + key.getCount());
52.        context.write(text, NullWritable.get());
53.    }
54. }
55. }
```

步骤 04 运行项目。

现在运行项目，在 IDEA 的 run > Edit Configuratin 中添加两个参数，如图 4-5 所示。由于代码是在 IDEA 中直接运行，即运行在本地，所以传递本地的目录即可，如果运行在 Hadoop 集群中，请传递 HDFS 上的文件和目录。

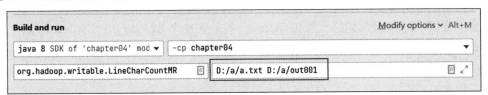

图 4-5

运行完成后查看输出的结果，如下所示：

```
This Example show how    21
to implements Writable   22
Hello This Is First Line .    26
```

从上面输出的结果可以看出，已经按每行字符数量进行了排序，而这是 comparaTo 的功能。这个例子演示了如何自定义 Writable 实现序列化的功能。

4.4 Partitioner 分区编程

Partitioner 分区编程的主要功能是将不同的分类输出到不同的文件中。这样在查询数据时，可以根据某个规定的类型查询相关的数据。

使用 Partitioner 分区必须给 Job 设置以下两个参数。

设置 Reducer 的数量，默认为 1：

```
job.setNumReduceTasks(2);
```

设置 Paratiner 类：

```
job.setPartitionerClass(Xxxx.class);
```

Map 的结果会通过 Partition 分发到 Reducer 上，Reducer 做完 Reduce 操作后，通过 OutputFormat 输出结果，如图 4-6 所示。

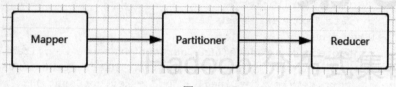

图 4-6

Mapper 最终处理的键值对<key, value>，需要送到 Reducer 中去合并，合并的时候，有相同 key 的键值对会输送到同一个 Reducer 上。哪个 key 到哪个 Reducer 的分配过程，是由 Partitioner 规定的。它只有一种方法：

```
getPartition(Text key, Text value, int numPartitions)
```

输入参数是 Map 的结果对<key, value>和 Reducer 的数目，输出则是分配的 Reducer（整数编号）的数量。就是指定 Mapper 输出的键值对到哪一个 Reducer 上去。系统默认的 Partitioner 是 HashPartitioner，它以 Key 的 Hash 值对 Reducer 的数目取模，得到对应的 Reducer。这样保证如果有相同的 Key 值，肯定会被分配到同一个 Reducer 上。如果有 N 个 Reducer，编号就为 0,1,2,3,…,(N-1)。

在执行 job 之前，设置 Reduce 的个数：

```
job.setNumReduceTasks(2);
```

则默认会根据 Key 的 Hash 值，将数据分别输出到 2 个文件中。默认使用 HashPartitioner 类源代码如下：

```
@InterfaceAudience.Public
@InterfaceStability.Stable
public class HashPartitioner<K2, V2> implements Partitioner<K2, V2> {
   public void configure(JobConf job) {}
   /** Use {@link Object#hashCode()} to partition. */
   public int getPartition(K2 key, V2 value,
                     int numReduceTasks) {
      return (key.hashCode() & Integer.MAX_VALUE) % numReduceTasks;
   }
}
```

通过上面的代码算法可知：key.hashCode & Integer.MAX_VALUE 是将数据转成正数，然后与设置 Reduce 的个数取模。

现在先来做一个测试，在【代码 4-4】中，我们先修改 Reducer 的个数，给 job 添加 Reducer 个数的代码如下，其中"…"为前后省略的代码。

```
...
FileOutputFormat.setOutputPath(job, dest);
//设置 Reducer 的个数
job.setNumReduceTasks(2);
boolean b = job.waitForCompletion(true);
...
```

然后再重新执行【代码 4-4】，查看输出结果，输出的文件成为两个，且这两个文件中的内容不会相同。

```
part-r-00000
part-r-00001
```

现在我们可以自己开发 Partitioner，修改默认使用 hashCode 这个规则。比如我们根据字符的个数进行输出，将字符个数大于 25 的输出到一个文件中，小于等于 25 个字符的输出到另一个文件中。现在开发 Partitioner 类的继承类如【代码 4-5】所示。第 7 行代码通过判断每一行的字符个数，将数据保存到不同的文件中，通过返回 int 的值，可以区分保存的文件。

【代码 4-5】LineCharCountPartitioner.java

```
1.  package org.hadoop.writable;
2.  import org.apache.hadoop.io.NullWritable;
3.  import org.apache.hadoop.mapreduce.Partitioner;
4.  public class LineCharCountPartitioner extends Partitioner<LineCharCountWritable, NullWritable> {
5.      @Override
6.      public int getPartition(LineCharCountWritable key, NullWritable value, int i) {
7.          if(key.getCount()>25){
8.              return 1;
9.          }else{
10.             return 0;
11.         }
12.     }
13. }
```

然后将自定义的 Partitioner 设置到 job 中去即可：

```
    job.setNumReduceTasks(2);
job.setPartitionerClass(LineCharCountPartitioner.class);
```

运行并查看结果，发现 25 个字符以下的数据输入到 part-r-00000 文件中，25 个字符以上的数据输出到 part-r-00001 文件中去了。

4.5 自定义排序

在 Hadoop 开发中，通过 Mapper 输出给 Reducer 的数据已经根据 key 值进行了排序，即 Mapper<Key1_input,Value1_input,Key2_out,Value2_out>，其中默认根据 key2_out 进行排序。

所以，如果希望输出的数据可以实现排序，可以通过以下方式实现：

- 开发 JavaBean 实现接口 WritableComparable，此类是 Writable 的子类。
- 将 JavaBean 作为 Mapper 的 Key2 输出给 Reducer。
- 实现 JavaBean 的 comparaTo 方法，实现比较。

在开发时，可以开发多个 MapReduce 处理过程，形成一个处理链，后面的 MapReduce 处理前面 MapReduce 输出的结果，直接到满足最后所需要的数据格式。

以下开发一个自定义排序的 Wriable 类，为大家演示自定义排序的开发过程。

1. 实现 WritableComparable 接口

之前 WordCount 输出的数据，默认是升序排序的。现在我们可以自定义一个排序规则，让它按倒序排序。

【代码 4-6】MyText.java

```java
1.  package org.hadoop.sort;
2.  public class MyText implements WritableComparable<MyText> {
3.      private String text;
4.      @Override
5.      public int compareTo(MyText o) {
6.          return o.text.compareTo(this.text);
7.      }
8.      @Override
9.      public void write(DataOutput out) throws IOException {
10.         out.writeUTF(text);
11.     }
12.     @Override
13.     public void readFields(DataInput in) throws IOException {
14.         text = in.readUTF();
15.     }
16.     public String getText() {
17.         return text;
18.     }
19.     public void setText(String text) {
20.         this.text = text;
21.     }
22. }
```

然后将 MyText 作为 Mapper 的输出 Key 值。首先设置 job 输出的格式：

```
    job.setMapOutputKeyClass(MyText.class);
job.setMapOutputValueClass(LongWritable.class);
```

设置为 Mapper 的输出：

```
    public static class WordCountMapper extends Mapper<LongWritable, Text, MyText, LongWritable> { ... }
```

设置 Reducer 接收 MyText 作为输入：

```
public static class WordCountReducer extends Reducer<MyText, LongWritable, Te
xt, LongWritable> {  .. }
```

执行完成以后，查看输出的结果。

```
mrchiian       1
value.toString();    1
throws    4
text.set(s);    1
```

由结果可以看出，已经是倒序了。

2. 实现 WritableComparator 接口

如果不希望修改 Mapper 输出的 Key 值部分，则可以定义一个 Comparator，如【代码 4-7】所示。

【代码 4-7】MyComparator.java

```
1.   package org.hadoop.sort;
2.   public class MyComparator extends WritableComparator {
3.       public MyComparator() {
4.           //注意构造部分，第二个的 true 必须传递，否则会抛出异常
5.           super(Text.class,true);
6.       }
7.       @Override
8.       public int compare(WritableComparable a, WritableComparable b) {
9.           int c = b.compareTo(a);
10.          return c;
11.      }
12.  }
```

然后在 job 中添加一个排序对象即可：

```
job.setSortComparatorClass(MyComparator.class);
```

最后，执行并查看结果，已经可以按指定的顺序显示了。

4.6 Combiner 编程

Combiner 指的是在 Mapper 以后先对数据进行一个计算，然后再将数据发送到 Reducer。开发 Combiner 就是开发一个继承 Reducer 的类以实现 Reducer 中相同的功能。Combiner 是可插拔的组件。但仅限于 Combiner 的业务与 Reducer 的业务相同的情况下使用。在开发中，可以通过调用 job.setCombinerClass(..)添加 Combiner 类。

以下过程说明没有 Combiner 和有 Combiner 的两种不同的执行过程：

源数据如下：

```
Hello    Jack
Hello    Mary
Hello    Jack
Hello    Jack
```

没有 Combiner 的情况：

```
Mapper              Shuffle              Reducer
<Hello,1>           <Hello,[1,1,1,1]>    <Hello,4>
<Jack,1>            <Jack,[1,1,1]>       <Jack,3>
<Hello,1>           <Mary,[1]>           <Mary,1>
<Mary,1>
<Hello,1>
<Jack,1>
<Hello,1>
<Jack,1>
```

有 Combiner 的情况：

```
Mapper              Combiner            Shuffle             Reducer
<Hello,1>           <Hello,4>           <Hello,[4]>         <Hello,4>
<Jack,1>            <Jack,3>            <Jack,[3]>          <Jack,3>
<Hello,1>           <Mary,1>            <Mary,[1]>          <Mary,1>
<Mary,1>
<Hello,1>
<Jack,1>
<Hello,1>
<Jack,1>
```

通过上面的结果可以看到，在 Combiner 中，已经对数据进行了一次合并计算操作。这将减少在最后的 Reducer 中遍历处理的次数。

现在添加这个 Combiner，在启动之前调用 setCombinerClass(..)即可。

```
//设置一个 Combiner
job.setCombinerClass(PhoneDataReducer.class);
//开始任务
job.waitForCompletion(true);
```

由于 Combiner 类就是 Reducer 类的子类，所以此处就不再赘述。读者可以直接将 Reducer 类作为 Combiner 设置到 job 中去即可。

4.7 默认 Mapper 和默认 Reducer

如果没有给 Job 设置 Mapper 和 Reducer，则将会使用默认的 Mapper 和 Reducer。默认的 Mapper 和 Reducer 什么都不会做，只是进行文件的快速复制，如【代码 4-8】所示。此代码中，并没有设置 Mapper 和 Reducer 类，将会使用默认的 Mapper 和 Reducer 类。

【代码 4-8】DefaultMR.java

```
1.    package org.hadoop.dft;
2.    public class DefaultMR extends Configured implements Tool {
3.        @Override
4.        public int run(String[] args) throws Exception {
5.            if(args.length!=2){
```

```
6.              System.out.println("usage : <in> <out>");
7.              return -1;
8.          }
9.          Configuration conf = getConf();
10.         Job job = Job.getInstance(conf,"DefaultMR");
11.         job.setJarByClass(this.getClass());
12.         FileSystem fs = FileSystem.get(conf);
13.         if(fs.exists(new Path(args[1]))){
14.             fs.delete(new Path(args[1]),true);
15.         }
16.         FileInputFormat.setInputPaths(job,new Path(args[0]));
17.         FileOutputFormat.setOutputPath(job,new Path(args[1]));
18.         int code = job.waitForCompletion(true)?0:1;
19.         return code;
20.     }
21.     public static void main(String[] args) throws Exception {
22.         int code = ToolRunner.run(new DefaultMR(),args);
23.         System.exit(code);
24.     }
25. }
26.
```

4.8 倒排索引

倒排索引用于统计并记录某个单词在一个文件中出现的次数及位置。我们可以实现一个简单的算法，统计单词在一个文件中出现的次数。假如存在以下两个文件：

a.txt 文件中的内容为：

```
Hello Jack
Hello Jack
```

b.txt 文件中的内容为：

```
Hello Mary
```

则统计完成以后的结果为：

单词	文件	出现次数	文件	出现次数	总出现次数
Hello	a.txt,	2	b.txt,	1	3
Jack	a.txt,	2			2
Mary			b.txt,	1	1

处理的思路可以是先根据 Word+文件名做一次统计，结果为：

单词	文件	出现次数
Hello	a.txt	2
Hello	b.txt	1
Jack	a.txt	2
Mary	b.txt	1

然后再对上面的结果进行处理，以 Word 为 key 以文件名、次数为 Value 进行再处理，并最终输出要求的结果。

第一个 MapReduce 程序，用于将两个文件中的数据先按单词+\t+文件+\t+出现次数进行统计。

【代码 4-9】InverseMR1.java

```
1.  package org.hadoop.invise;
2.  public class InverseMR1 extends Configured implements Tool {
3.      @Override
4.      public int run(String[] args) throws Exception {
5.          Configuration conf = getConf();
6.          FileSystem fs = FileSystem.get(conf);
7.          Path dest = new Path("D:/a/out002");
8.          if (fs.exists(dest)) {
9.              fs.delete(dest, true);
10.         }
11.         Job job = Job.getInstance(conf, "InverseIndex");
12.         job.setJarByClass(getClass());
13.         job.setMapperClass(IIMapper.class);
14.         job.setMapOutputKeyClass(Text.class);
15.         job.setMapOutputValueClass(LongWritable.class);
16.         job.setReducerClass(IIReducer.class);
17.         job.setOutputKeyClass(Text.class);
18.         job.setOutputValueClass(NullWritable.class);
19.         job.setInputFormatClass(TextInputFormat.class);
20.         job.setOutputFormatClass(TextOutputFormat.class);
21.         FileInputFormat.setInputPaths(job, new Path("D:/a/in"));
22.         FileOutputFormat.setOutputPath(job, dest);
23.         int code = job.waitForCompletion(true) ? 0 : 1;
24.         return code;
25.     }
26.     public static class IIMapper extends Mapper<LongWritable, Text, Text, LongWritable> {
27.         private String fileName = "";
28.         private Text key = new Text();
29.         private LongWritable value = new LongWritable(0L);
30.         @Override
31.         public void map(LongWritable key, Text value, Context context) throws IOException, InterruptedException {
32.             String[] strs = value.toString().split("\\s+");
33.             for (String str : strs) {
34.                 this.key.set(str + "\t" + fileName);
35.                 this.value.set(1L);
36.                 context.write(this.key, this.value);
37.             }
38.         }
39.         @Override
40.         protected void setup(Context context) throws IOException, InterruptedException {
41.             InputSplit split = context.getInputSplit();
```

```
42.            if (split instanceof FileSplit) {
43.                FileSplit fileSplit = (FileSplit) split;
44.                fileName = fileSplit.getPath().getName();
45.            }
46.        }
47.    }
48.    public static class IIReducer extends Reducer<Text, LongWritable, Text, NullWritable> {
49.        @Override
50.        public void reduce(Text key, Iterable<LongWritable> values, Context context) throws IOException, InterruptedException {
51.            long sum = 0L;
52.            for (LongWritable l : values) {
53.                sum += l.get();
54.            }
55.            key.set(key.toString() + "\t" + sum);
56.            context.write(key, NullWritable.get());
57.        }
58.    }
59.    public static void main(String[] args) throws Exception {
60.        int code = ToolRunner.run(new InverseMR1(), args);
61.        System.exit(code);
62.    }
63. }
```

统计后的结果如下：

```
Hello   a.txt   2
Hello   b.txt   1
Jack    a.txt   2
Mary    b.txt   1
```

第二个 MapReduce 程序，用于将上面的结果再根据单词进行统计，如【代码 4-10】所示。

【代码 4-10】InverseMR2.java

```
1.  package org.hadoop.inverse;
2.  public class InverseMR2 extends Configured implements Tool {
3.      @Override
4.      public int run(String[] args) throws Exception {
5.          Configuration conf = getConf();
6.          FileSystem fs = FileSystem.get(conf);
7.          Path dest = new Path("D:/a/out003");
8.          if (fs.exists(dest)) {
9.              fs.delete(dest, true);
10.         }
11.         Job job = Job.getInstance(conf, "InverseIndex2");
12.         job.setJarByClass(getClass());
13.         job.setMapperClass(IIMapper2.class);
14.         job.setMapOutputKeyClass(Text.class);
15.         job.setMapOutputValueClass(Text.class);
16.         job.setReducerClass(IIReducer2.class);
```

```
17.         job.setOutputKeyClass(Text.class);
18.         job.setOutputValueClass(LongWritable.class);
19.         job.setInputFormatClass(TextInputFormat.class);
20.         job.setOutputFormatClass(TextOutputFormat.class);
21.         FileInputFormat.setInputPaths(job, new Path("D:/a/out002"));
22.         FileOutputFormat.setOutputPath(job, dest);
23.         int code = job.waitForCompletion(true) ? 0 : 1;
24.         return code;
25.     }
26.     public static class IIMapper2 extends Mapper<LongWritable, Text, Text, Text> {
27.         private Text key = new Text();
28.         private Text value = new Text();
29.         @Override
30.         public void map(LongWritable key, Text value, Context context) throws IOException, InterruptedException {
31.             String[] strs = value.toString().split("\\s+");
32.             this.key.set(strs[0]);//Hello
33.             this.value.set(strs[1] + "\t" + strs[2]);//a.txt,1
34.             context.write(this.key, this.value);
35.         }
36.     }
37.     public static class IIReducer2 extends Reducer<Text, Text, Text, LongWritable> {
38.         private LongWritable sum = new LongWritable(0L);
39.         @Override
40.         public void reduce(Text key, Iterable<Text> values, Context context) throws IOException, InterruptedException {
41.             this.sum.set(0L);
42.             String str = "";
43.             for (Text t : values) {
44.                 String[] strs = t.toString().split("\t");
45.                 this.sum.set(this.sum.get() + Long.parseLong(strs[1]));
46.                 str += "\t" + t.toString();
47.             }
48.             key.set(key.toString() + "\t" + str);
49.             context.write(key, this.sum);
50.         }
51.     }
52.     public static void main(String[] args) throws Exception {
53.         int code = ToolRunner.run(new InverseMR2(), args);
54.         System.exit(code);
55.     }
56. }
```

执行后的结果如下：

```
Hello      b.txt   1    a.txt   2    3
Jack       a.txt   2    2
Mary       b.txt   1    1
```

4.9 Shuffle

Shuffle 是 MapReduce 的核心。Shuffle 的本义是洗牌、混洗，把一组有一定规则的数据尽量转换成一组无规则的数据，越随机越好。MapReduce 中的 Shuffle 更像是洗牌的逆过程，把一组无规则的数据尽量转换成一组具有一定规则的数据。

为什么 MapReduce 计算模型需要 Shuffle 过程？我们都知道 MapReduce 计算模型一般包括两个重要的阶段：Map 是映射，负责数据的过滤分发；Reduce 是规约，负责数据的计算归并。Reduce 的数据来源于 Map，Map 的输出即是 Reduce 的输入，Reduce 需要通过 Shuffle 来获取数据。

从 Map 输出到 Reduce 输入的整个过程可以广义地称为 Shuffle。Shuffle 横跨 Map 端和 Reduce 端，在 Map 端包括 Spill 过程，在 Reduce 端包括 copy 和 sort 过程，如图 4-7 所示。

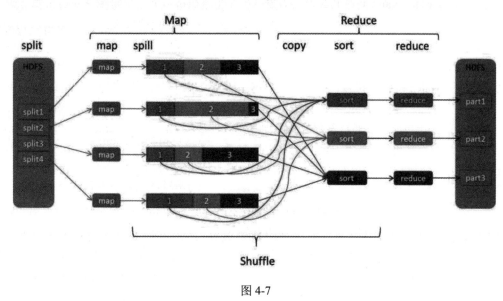

图 4-7

4.9.1 Spill 过程

Spill 过程包括输出、排序、溢写、合并等步骤，如图 4-8 所示。

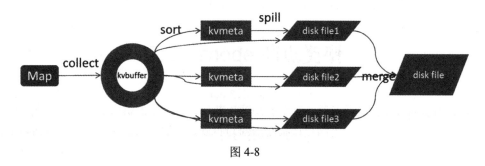

图 4-8

Spill 的第一个阶段 Collect，每个 Map 任务不断地以<key, value="">对的形式把数据输出到内存中，以构造的一个环形数据结构。使用环形数据结构是为了更有效地使用内存空间，在内存中放置尽可能多的数据。

这个数据结构其实就是个字节数组，叫 Kvbuffer，名如其义，但是这里面不光放置了<key, value="">数据，还放置了一些索引数据，给放置索引数据的区域起了一个 Kvmeta 的别名，在 Kvbuffer 的一块区域上穿了一个 IntBuffer（字节序采用的是平台自身的字节序）的马甲。<key, value="">数据区域和索引数据区域在 Kvbuffer 中是相邻不重叠的两个区域，用一个分界点来划分两者，这个分界点不是亘古不变的，而是每次 Spill 之后都会更新一次。初始的分界点是 0，<key, value="">数据的存储方向是向上增长，索引数据的存储方向是向下增长。

Kvbuffer 的存放指针 bufindex 是一直闷着头地向上增长，比如 bufindex 初始值为 0，一个 Int 型的 key 写完之后，bufindex 增长为 4，一个 Int 型的 value 写完之后，bufindex 增长为 8。

索引是对<key, value="">在 kvbuffer 中的索引，是个四元组，包括：value 的起始位置、key 的起始位置、partition 值、value 的长度，占用四个 Int 长度，Kvmeta 的存放指针 Kvindex 每次都是向下跳四个"格子"，然后再向上一个格子一个格子地填充四元组的数据。比如，Kvindex 初始位置是-4，当第一个<key, value="">写完之后，(Kvindex+0)的位置存放 value 的起始位置，(Kvindex+1)的位置存放 key 的起始位置，(Kvindex+2)的位置存放 partition 的值，(Kvindex+3)的位置存放 value 的长度，然后 Kvindex 跳到-8 位置，等第二个<key, value="">和索引写完之后，Kvindex 跳到-32 位置。

Kvbuffer 的大小虽然可以通过参数设置，但是总共就那么大，<key, value="">和索引不断地增加，Kvbuffer 总有不够用的那一天，那时怎么办？把数据从内存刷到磁盘上再接着往内存写数据，把 Kvbuffer 中的数据刷到磁盘上的过程就叫 Spill，内存中的数据满了就自动地 Spill 到具有更大空间的磁盘上。

关于 Spill 触发的条件，也就是 Kvbuffer 用到什么程度开始 Spill，还是要讲究一下的。如果把 Kvbuffer 用得比较满，一点缝隙都不剩的时候再开始 Spill，那么 Map 任务就需要等 Spill 腾出空间之后才能继续写数据；如果 Kvbuffer 只是满到一定程度，比如 80%的时候就开始 Spill，那在 Spill 的同时，Map 任务还能继续写数据，如果 Spill 够快，Map 可能都不需要为空闲空间而发愁。两利相衡取其大，一般选择后者。

Spill 这个重要的过程由 Spill 线程承担，Spill 线程从 Map 任务接到"命令"之后就开始正式工作，这就叫 SortAndSpill，原来不仅仅是 Spill，在 Spill 之前还有个颇具争议性的 Sort。

4.9.2　Sort 过程

Sort 先把 Kvbuffer 中的数据按照 partition 值和 key 两个关键字升序排序，移动的只是索引数据，排序结果是 Kvmeta 中数据按照 partition 为单位聚集在一起，同一 partition 内的按照 key 有序排列。

Spill 线程为这次 Spill 过程创建一个磁盘文件：从所有的本地目录中轮训查找能存储这么大空间的目录，找到之后在其中创建一个类似于"spill12.out"的文件。Spill 线程根据排过序的 Kvmeta 挨个 partition 的把<key, value="">数据遍历到这个文件中，一个 partition 对应的数据遍历完之后顺序地遍历下个 partition，直到把所有的 partition 遍历完。一个 partition 在文件中对应的数据也叫段（segment）。

所有的 partition 对应的数据都放在这个文件里，虽然是顺序存放的，但是怎么直接知道某个 partition 在这个文件中存放的起始位置呢？强大的索引又出场了。有一个三元组记录某个 partition

对应的数据在这个文件中的索引：起始位置、原始数据长度、压缩之后的数据长度，一个 partition 对应一个三元组。然后把这些索引信息存放在内存中，如果内存中放不下了，后续的索引信息就需要写到磁盘文件中了：从所有的本地目录中轮训查找能存储这么大空间的目录，找到之后在其中创建一个类似于"spill12.out.index"的文件，文件中不光存储了索引数据，还存储了 crc32 的校验数据。

注意：spill12.out.index 不一定在磁盘上创建，如果内存（默认 1MB 空间）中能放得下就放在内存中，即使在磁盘上创建了，它和 spill12.out 文件也不一定在同一个目录下。

每一次 Spill 过程就会最少生成一个 out 文件，有时还会生成 index 文件，Spill 的次数也烙印在文件名中。索引文件和数据文件的对应关系如图 4-9 所示。

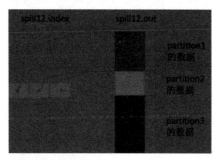

图 4-9

在 Spill 线程如火如荼地进行 SortAndSpill 工作的同时，Map 任务不会因此而休息，而是一如既往地进行着数据输出。Map 还是把数据写到 kvbuffer 中，问题是：<key, value="">只按照 bufindex 指针向上增长，kvmeta 按照 Kvindex 向下增长，是保持指针起始位置不变呢，还是另谋它路？如果保持指针起始位置不变，很快 bufindex 和 Kvindex 就相撞了，相撞之后再重新开始或者移动内存都比较麻烦，这很不可取。Map 取 kvbuffer 中剩余空间的中间位置，用这个位置设置为新的分界点，bufindex 指针移动到这个分界点，Kvindex 移动到这个分界点的-16 位置，然后两者就可以和谐地按照自己既定的轨迹放置数据了，当 Spill 完成，空间腾出之后，不需要做任何改动继续前进。

Map 任务总要把输出的数据写到磁盘上，即使输出数据量很小在内存中全部能装得下，在最后也会把数据刷到磁盘上。

4.9.3 Merge 过程

Map 任务如果输出数据量很大，可能会进行好几次 Spill，out 文件和 Index 文件会产生很多，分布在不同的磁盘上。最后把这些文件进行合并的 merge 过程闪亮登场。

Merge 过程如何知道产生的 Spill 文件都在哪了呢？从所有的本地目录上扫描得到产生的 Spill 文件，然后把路径存储在一个数组中。Merge 过程又怎么知道 Spill 的索引信息呢？没错，也是从所有的本地目录上扫描得到 Index 文件，然后把索引信息存储在一个列表里。到这里，又遇到了一个值得深思的地方。在之前 Spill 过程中的时候，为什么不直接把这些信息存储在内存中呢，何必又多了这步扫描的操作？特别是 Spill 的索引数据，之前当内存超限之后就把数据写到磁盘，现在又要从磁盘把这些数据读出来，还是需要装到更多的内存中。之所以多此一举，是因为这时 kvbuffer 这个内存大户已经不再使用，可以回收，有内存空间来装这些数据了。然后为 merge 过程创建一个叫

file.out 的文件和一个叫 file.out.Index 的文件用来存储最终的输出和索引。partition 是一个接一个地进行合并输出。对于某个 partition 来说，从索引列表中查询这个 partition 对应的所有索引信息，每个对应一个段插入到段列表中。也就是这个 partition 对应一个段列表，记录所有的 Spill 文件中对应的这个 partition 那段数据的文件名、起始位置、长度，等等。

然后对这个 partition 对应的所有的 segment 进行合并，目标是合并成一个 segment。当这个 partition 对应很多个 segment 时，会分批地进行合并：先从 segment 列表中把第一批取出来，以 key 为关键字放置成最小堆，然后从最小堆中每次取出最小的<key, value="">输出到一个临时文件中，这样就把这一批段合并成一个临时的段，把它加回到 segment 列表中；再从 segment 列表中把第二批取出来合并输出到一个临时 segment，把其加入到列表中；这样往复执行，直到剩下的段是一批，输出到最终的文件中。

1. Copy

Reduce 任务通过 HTTP 向各个 Map 任务拖取它所需要的数据。每个节点都会启动一个常驻的 HTTP Server，其中一项服务就是响应 Reduce 拖取 Map 数据。当有 MapOutput 的 HTTP 请求过来的时候，HTTP Server 就读取相应的 Map 输出文件中对应这个 Reduce 部分的数据，通过网络流输出给 Reduce。

Reduce 任务拖取某个 Map 对应的数据，如果在内存中能放得下这次数据的话，就直接把数据写到内存中。Reduce 要向每个 Map 去拖取数据，在内存中每个 Map 对应一块数据，当内存中存储的 Map 数据占用空间达到一定程度的时候，开始启动内存中的数据 Merge，把内存中的数据 Merge 输出到磁盘上一个文件中。

如果在内存中不能放得下这个 Map 的数据的话，直接把 Map 数据写到磁盘上，在本地目录创建一个文件，从 HTTP 流中读取数据然后写到磁盘，使用的缓存区大小是 64KB。拖一个 Map 数据过来就会创建一个文件，当文件数量达到一定阈值时，开始启动磁盘文件 Merge，把这些文件合并输出到一个文件中。

有些 Map 的数据较小可以放在内存中，有些 Map 的数据较大需要放在磁盘上，这样最后 Reduce 任务拖过来的数据有些放在内存中、有些放在磁盘上，最后会对这些来一个全局合并。

2. Merge Sort

这里使用的 Merge 和 Map 端使用的 Merge 过程一样。Map 的输出数据已经是有序的，Merge 进行一次合并排序，所谓 Reduce 端的 sort 过程就是这个合并的过程。一般 Reduce 是一边复制一边 sort，即复制和 sort 两个阶段是重叠而不是完全分开的。

Reduce 端的 Shuffle 过程至此结束。

4.10 本章小结

本章主要讲解了以下内容：

- MapReduce 的过程被显式分为两部分：Mapper 和 Reducer。

- 在 MapReduce 中被传递的类必须实现 Hadoop 的序列化接口 Writable。
- Mapper 中输出的 key 可以实现排序，此类必须实现接口 WritableComparable 接口。
- 多个 MapReduce 可以分别打成不同的 Jar 包执行。这样后面的 MapReduce 可以使用前面 MapReduce 输出的数据。
- Partitioner 编程用于将不同规则的数据输出到指定编号的 Reduce 上去。
- Combiner 类似于一个前置的 Reduce，用于在将数据输出到 Reduce 之前进行一次数据的合并操作。
- MapReduce 开发过程归纳如下：
  - 研究业务需要输出的格式。
  - 自定义一个类继承 Mapper 类，并指定输入输出的格式。
  - 重写 map 方法实现具体的业务逻辑。注意使用 context 执行输出操作。
  - 自定义一个类继承 Reduce 类，并指定输入输出格式。
  - 重写 reduce 方法，实现自己的业务代码。注意使用 context 执行输出操作。
  - 开发一个 main 方法，通过 Job 对象进行组装。
  - 打成 jar 包，指定主类，发送到 Linux 上，通过 yarn jar 命令来启动 MapReduce。
- MapReduce 的执行流程归纳如下：
  - Client 通过 RPC 将请求提交给 ResouceManager。
  - ResouceManager 在接收到请求以后返回一个 jobID 给 Client。
  - Client 将 jar 包上传到 HDFS（默认 HDFS 保存 10 份），且程序执行完成以后删除 jar 包文件。
  - Client 通知 ResourceManager 保存数据的描述信息。
  - ResouceManager 通过公平调度开启任务，将任务放到任务调度队列。
  - NodeManager 通过心跳机制向 ResourceManager 领取任务。
  - NodeManager 向 HDFS 领取所需要的 jar 包。开始执行任务。
- Partitioner 分区编程，就是控制将不同的结果输出到指定的文件中。
- 开启一个 Reducer 就会拥有一个输出文件。通过设置 job.setNumReduceTask 可以设置开启 Reducer 的个数，此数值必须大于等于 Partitioner 分区返回的数量。
- 当启动的 Reducer 大于 Partitioner 返回的数量时，将会生成一些空的文件。当 Reducer 数量小于 Partitioner 返回的数量时，将直接抛出异常 Illege Partition counter error。所以，在设置 Reducer 数量时，应该考虑 Partitioner 返回的分区的个数。
- Mapper 在输出数据给 Reducer 时，会根据不同的分区编号分发到不同的 Reducer 上去，每一个 Reducer 都会有一个编号。

第 5 章

Hadoop 输入输出

本章主要内容：

- InputFormat 输入类。
  - ➢ TextInputFormat 默认输入类。
  - ➢ 自定义 InputFormat 读取文本文件中的内容。
  - ➢ 自定义 InputFormat 读取数据库中的内容。
  - ➢ 读取顺序文件。
- OutputFormat 输出类。
  - ➢ TextOutputFormat 默认输出类。
  - ➢ 自定义 OutputFormat 输出到文本文件。
  - ➢ 自定义 OutputFormat 输出到数据库中。
  - ➢ 自定义 OutputFormat 将数据保存到 HBase 中。
  - ➢ 输出顺序文件。
  - ➢ 压缩输出文件。

在前面章节的代码中，输入源都是文本文件，在具体的业务中，文本文件确实占的比重比较多，如日志文件都是文本文件类型。但也有可能需要处理其他类型的输入源，如输入（输出）源为数据库、输入源来自于网络等。

默认情况下，Hadoop 解析 txt 文件返回的数据格式为：<LongWritable,Text>，即前一个值为字节偏移量，后面的 Text 为每一次读取的一行数据。

在 Hadoop 中，所有的输入流都应该是 InputFormat 的子类，而所有的输出流都应该是 OutputFormat 的子类。InputFormat<T1,T2>接收两个泛型，其中 T1 表示返回的 Key 类型，T2 表示返回的 Value 类型，这些返回值将作为 Mapper 类的接收泛型。OutputFormat<T1,T2>也接收两个泛型，其中 T1 表示输出的 Key 类型，而 T2 为输出的 Value 类型，此两个类型，应该是 Reducer 最终输出的类型。

现在我们先来练习开发自定义 InputFormat 类，让 LongWritable 为行号而不是字节的偏移量，此时，就应该通过继承 InputFormat 实现自定义文件输入流。

5.1 自定义文件输入流

FileInputFormat 默认使用 TextFileInputFormat 类，此类默认读取文本文件，Key 为 LongWritable 类型的字节偏移量，Value 为 Text 类型，表示这一行的数据。现在我们尝试重新定义一个读取文本文件的类，Key 为 LongWritable 类型的行号（注意是行号，不是字节偏移量），Value 是这一行的文本值与之前相同。

5.1.1 自定义 LineTextInputFormat

自定义的 LineTextInputFormat 用于读取文本文件，且返回的 Key 为 Int 类型的行号，数据为 Text 类型的值。

InputFormat 为所有输入流的父类。而 FileInputFormat 为所有文件类型输入流的父类。所以，这里我们可以直接继承 FileInputFormat 即可。它们的关系如图 5-1 所示。

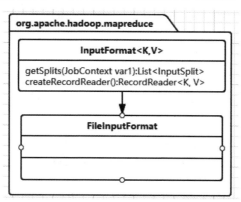

图 5-1

以下自定义 InputFormat 从文本文件中读取数据，返回<行号,数据>格式，即返回的 key 为 IntWritable 类型的行号和 Text 类型的行数据。在自定义的 LineNumberInputFormat 类中，最重要的就是开发一个实现 RecordReader 的子类。此类是真正实现文件读取的具体类。

【代码 5-1】LineNumberInputFormat.java

```
1. package org.hadoop.input;
2. public class LineNumberInputFormat extends FileInputFormat<LongWritable, Text> {
3.     @Override
4.     public RecordReader<LongWritable, Text> createRecordReader(InputSplit split, TaskAttemptContext context)
5.             throws IOException, InterruptedException {
6.         RecordReader<LongWritable, Text> reader = new LineTextReader();
```

```java
7.         reader.initialize(split, context);
8.         return reader;
9.     }
10.    public static class LineTextReader extends RecordReader<LongWritable, Text> {
11.        private Long fileSize;//总大小，用于计算这个块的读取比例
12.        private Long sum = 0L;//读取到多少字节了，可与fileSize计算出读取比例
13.        private BufferedReader br;//读取一次读取一行IO对象
14.        private LongWritable line = new LongWritable(0L);//行号
15.        private Text value = new Text();//行数据
16.        @Override
17.        public void initialize(InputSplit split, TaskAttemptContext context)
18.                throws IOException, InterruptedException {
19.            if (split instanceof FileSplit) {//判断是否是文件类型
20.                FileSplit fileSplit = (FileSplit) split;
21.                fileSize = fileSplit.getLength();
22.                Path path = fileSplit.getPath();
23.                FileSystem fs = FileSystem.get(context.getConfiguration());
24.                br = new BufferedReader(new InputStreamReader(fs.open(path)));
25.            }
26.        }
27.        /**
28.         * 用于返回是否还有下一行数据
29.         */
30.        @Override
31.        public boolean nextKeyValue() throws IOException, InterruptedException {
32.            String line = br.readLine();
33.            if (line != null) {
34.                this.line.set(this.line.get() + 1);
35.                value.set(line);
36.                sum += line.getBytes().length;//读取到多少数据了
37.                return true;
38.            }
39.            return false;
40.        }
41.        /**
42.         * 获取当前Key值，只有nextKeyValue返回true时才会调用这个方法
43.         */
44.        @Override
45.        public LongWritable getCurrentKey() throws IOException, InterruptedException {
46.            return this.line;
47.        }
48.        /**
49.         * 获取当前行数据
50.         */
51.        @Override
```

```
52.         public Text getCurrentValue() throws IOException, InterruptedException {
53.             return value;
54.         }
55.         /**
56.          * 获取进度
57.          */
58.         @Override
59.         public float getProgress()throwsIOException, InterruptedException{
60.             return 1.0F * sum / fileSize;
61.         }
62.         /**
63.          * 最后关闭输入流
64.          */
65.         @Override
66.         public void close() throws IOException {
67.             if (br != null) {
68.                 br.close();
69.                 br = null;
70.             }
71.         }
72.     }
73. }
```

定义好自定义输入流后，在 main 方法中设置自定义文件输入流对象：

```
job.setInputFormatClass(LineNumberInputFormat.class);
```

以下 MR 程序，并没有设置 Mapper 和 Reducer，将会使用默认的 Mapper 和 Reducer，使用上述自定义的输入类，因为使用了默认的 Mapper 和 Reducer，所以会直接把从 LineNumberInputFormat 中读取的数据输出到指定的文件中去，以便于观察我们自定义的 LineNumberInputFormat 是否正确工作。

【代码 5-2】LineNumberMR.java

```
1. package org.hadoop.input;
2. /**
3.  * 使用自定义LineNumberTextInputFormat 读取文本文件，返回行号+行数据
4.  */
5. public class LineNumberMR extends Configured implements Tool {
6.     @Override
7.     public int run(String[] args) throws Exception {
8.         Configuration conf = getConf();
9.         FileSystem fs = FileSystem.get(conf);
10.        Path dest = new Path("D:/a/out001");
11.        if(fs.exists(dest)){
12.            fs.delete(dest,true);
13.        }
14.        Job job = Job.getInstance(conf,"LineNumberMR");
15.        job.setJarByClass(this.getClass());
16.        job.setOutputKeyClass(LongWritable.class);
```

```
17.         job.setOutputValueClass(Text.class);
18.         //设置输入流
19.         job.setInputFormatClass(LineNumberInputFormat.class);
20.         job.setOutputFormatClass(TextOutputFormat.class);
21.         FileInputFormat.setInputPaths(job,new Path("D:/a/a.txt"));
22.         FileOutputFormat.setOutputPath(job,dest);
23.         return job.waitForCompletion(true)?0:1;
24.     }
25.     public static void main(String[] args) throws Exception {
26.          int run = ToolRunner.run(new LineNumberMR(), args);
27.         System.exit(run);
28.     }
29.}
```

查看最后输出的结果，已经带有行号了：

```
1   Hello This Is First Line .
2   This Example show how
3   to implements Writable
```

5.1.2 自定义 ExcelInputFormat 类

ExcelInputFormat 用于读取 Excel 文件中的内容。读取 Excel 文件可以使用 Apache 的 POI 组件，所以需要添加 POI 的依赖：

```
<dependency>
    <groupId>org.apache.poi</groupId>
    <artifactId>poi</artifactId>
    <version>5.0.0</version>
</dependency>
```

由于在 Hadoop 的环境中并没有 Excel 的 POI 包，所以可以使用 maven-shade-plugin 将 POI 与开发代码打成一个完整的 Uber jar 包（可执行包），即将 POI 打到 jar 文件中去。这需要添加以下插件：

```
<build>
    <plugins>
        <plugin>
            <groupId>org.apache.maven.plugins</groupId>
            <artifactId>maven-shade-plugin</artifactId>
            <version>3.2.4</version>
            <executions>
                <execution>
                    <goals>
                        <goal>shade</goal>
                    </goals>
                </execution>
            </executions>
        </plugin>
    </plugins>
</build>
```

使用 POI 读取 Excel 文件的关键点在于读取 Excel 文件——HSSFWorkBook。以下是自定义 ExcelInputFormat 的源代码。

【代码 5-3】ExcelInputFormat.java

```java
1.   package org.hadoop.input.excel;
2.   /**
3.    * 读取 Excel 数据的 InputFormat<br>
4.    * 读取 Excel 文件中的所有 Sheet 然后再读取每一个 Sheet 中的所有行和列将数据转存成文本数据
5.    */
6.   public class ExcelInputFormat extends FileInputFormat<LongWritable, Text> {
7.       @Override
8.       public RecordReader<LongWritable, Text> createRecordReader(InputSplit split, TaskAttemptContext context)
9.               throws IOException, InterruptedException {
         //不需要调用 initialize 方法
10.          RecordReader<LongWritable, Text> rr = new ExcelRecordReader();
11.          return rr;
12.      }
13.      //RecorderReader 类的子类
14.      public static class ExcelRecordReader extends RecordReader<LongWritable, Text> {
15.          private HSSFWorkbook book;//表示 Excel 文件
16.          private int sheets = -1;//sheet 的个数
17.          private int currentSheetIndex = 0;//已经读取到当前 Sheet 的下标
18.          private HSSFSheet sheet;//当前 Sheet
19.          private int lastRowNum = -1;//每一个 Sheet 最后的行号
20.          private int currentRow = -1;//当前 Sheet 的当前行号
21.          private Text value = new Text();//这一行的数据
22.          @Override
23.          public void initialize(InputSplit split, TaskAttemptContext context)
24.                  throws IOException, InterruptedException {
25.              FileSplit fileSplit = (FileSplit) split;
26.              FileSystem fs = FileSystem.get(context.getConfiguration());
27.              InputStream in = fs.open(fileSplit.getPath());
28.              book = new HSSFWorkbook(in);
29.              sheets = book.getNumberOfSheets();//获取 Sheet 的个数
30.              if (sheets >= 1) {
31.                  sheet = book.getSheetAt(currentSheetIndex++);//获取第一个 Sheet
32.                  lastRowNum = sheet.getLastRowNum();//最后一行的行号
33.                  if (lastRowNum >= 0) {//如果有数据则将当前行先设置为 0
34.                      currentRow = 0;
35.                  }
36.              }
37.          }
38.          @Override
39.          public boolean nextKeyValue() throws IOException, InterruptedException {
```

```
40.            if (sheet != null && currentRow <= lastRowNum) {
41.                HSSFRow row = sheet.getRow(currentRow++);
42.                short cells = row.getLastCellNum();
43.                value.set("");
44.                for (int i = 0; i < cells; i++) {
45.                    HSSFCell cell = row.getCell(i);
46.                    String str = "";
47.                    if (cell.getCellType() == CellType.STRING) {
48.                        str = cell.getStringCellValue();
49.                    } else {
50.                        str = "" + cell.getNumericCellValue();
51.                    }
52.                    value.set(value.toString() + "|" + str);
53.                }
54.                return true;
55.            } else {//下一个 Sheet
56.                if (currentSheetIndex < sheets) {
                        //获取第一个 Sheet
57.                    sheet = book.getSheetAt(currentSheetIndex++);
58.                    if (sheet == null) {
59.                        return false;
60.                    }
61.                    lastRowNum = sheet.getLastRowNum();//最后一行的行号
62.                    if (lastRowNum >=0) {//如果有数据则将当前行先设置为 0
63.                        currentRow = 0;
64.                    }
65.                    //数据
66.                    if (currentRow <= lastRowNum) {//有数据
67.                        HSSFRow row = sheet.getRow(currentRow++);
68.                        if(row==null) {//已经没有数据了
69.                            return false;
70.                        }
71.                        short cells = row.getLastCellNum();
72.                        value.set("");
73.                        for (int i = 0; i < cells; i++) {
74.                            HSSFCell cell = row.getCell(i);
75.                            String str = "";
76.                            if (cell.getCellType() == CellType.STRING) {
77.                                str = cell.getStringCellValue();
78.                            } else {
79.                                str = "" + cell.getNumericCellValue();
80.                            }
81.                            value.set(value.toString() + "|" + str);
82.                        }
83.                        return true;
84.                    }
85.                }
86.                return false;
87.            }
88.        }
```

```
89.         @Override
90.         public LongWritable getCurrentKey() throws IOException, InterruptedException {
91.             return new LongWritable(currentRow);
92.         }
93.         @Override
94.         public Text getCurrentValue() throws IOException, InterruptedException {
95.             return value;
96.         }
97.         @Override
98.         public float getProgress() throws IOException, InterruptedException {
99.             return 0;
100.        }
101.        @Override
102.        public void close() throws IOException {
103.            if (book != null) {
104.                book.close();
105.            }
106.        }
107.    }
108. }
```

现在我们需要将 ExcelInputFormat 作为 Job 的输入流，只需要做：

```
job.setInputFormatClass(ExcelInputFormat.class);
```

【代码 5-4】 是 MapReduce 程序的完整源代码。

【代码 5-4】 ExcelInputMR.java

```
1.  package cn.hadoop.mapred;
2.  /**
3.   * 使用自定义的 ExcelInputFormat 类用于读取 Excel 文件
4.   * 不再定义 Mapper 和 Reducer 使用默认的就可以了
5.   */
6.  public class ExcelInputMR extends Configured implements Tool {
7.      @Override
8.      public int run(String[] args) throws Exception {
9.          if(args.length!=2) {
10.             System.out.println("Usage : <in> <out>");
11.             return -1;
12.         }
13.         Configuration conf = getConf();
14.         Path dest = new Path(args[1]);
15.         FileSystem fs = FileSystem.get(conf);
16.         if(fs.exists(dest)) {
17.             fs.delete(dest,true);
18.         }
19.         Job job = Job.getInstance(conf,"Excel");
20.         job.setJarByClass(getClass());
21.         job.setInputFormatClass(ExcelInputFormat.class);
```

```
22.            job.setOutputFormatClass(TextOutputFormat.class);
23.            //
24.            FileInputFormat.setInputPaths(job, new Path(args[0]));
25.            FileOutputFormat.setOutputPath(job, dest);
26.            int code = job.waitForCompletion(true)?0:1;
27.            return code;
28.        }
29.        public static void main(String[] args) throws Exception {
30.            int code = ToolRunner.run(new Demo06_ExcelInputFormatMR(), args);
31.            System.exit(code);
32.        }
33.  }
```

创建一个 Excel 文件，内容如下所示：

```
Jack    34        男
李四    54        女
```

执行 ExcelInputMR.java 并查看输出的结果，如下所示：

```
1       |Jack|34.0|男
2       |李四|54.0|女
```

5.1.3　DBInputFormat

从数据库中读取输入源，也是一种 InputFormat，Hadoop 已经为我们定义好了 DBInputFormat 类。DBInputFormat 可以用于读取数据库中输入的数据。以下是系统 DBInputFormat 类的部分源代码。

```
public class DBInputFormat<T extends DBWritable>
extends InputFormat<LongWritable, T> implements Configurable {
```

通过上述的源代码可知，DBInputFormat 为 InputFormat 的子类，且已经定义 Key 类型为 LongWritable。我们只需要定义 Value 类型即可，但 Value 类型必须是 DBWritable 的子类。因此，如果需要开发 InputFormat 读取数据库中的数据，则应该继承 DBInputFormat 且传递 Value 类型即可。

使用 DBInputFormat 需要根据连接数据库的类型，添加数据库的驱动依赖，本处连接的是 MySQL，所以添加 MySQL 的依赖如下：

```
<dependency>
    <groupId>mysql</groupId>
    <artifactId>mysql-connector-java</artifactId>
    <version>8.0.23</version>
</dependency>
```

首先我们需要创建一个数据库，以便于从表中读取数据。创建数据库、创建表的语句如下：

```
create database hadoop character set utf8mb4;
use hadoop;
create table stud(
    id int
    name varchar(128)
);
insert into stud values(1,'Jack'),(2,'Mary');
```

创建 DBWritable 的子类，如【代码 5-5】所示。实现 DBWritable 之后，需要多实现两个方法，分别用于通过 PreparedStatement 设置值和通过 ResultSet 读取值。

【代码 5-5】StudWritable.java

```
1.  package org.hadoop.input.db;
2.  public class StudWritable implements DBWritable, Writable {
3.      private Integer id;
4.      private String name;
5.      @Override
6.      public void write(DataOutput out) throws IOException {
7.          out.writeInt(id);
8.          out.writeUTF(name);
9.      }
10.     @Override
11.     public void readFields(DataInput in) throws IOException {
12.         id = in.readInt();
13.         name =in.readUTF();
14.     }
15.     @Override
16.     public void write(PreparedStatement statement) throws SQLException {
17.         statement.setInt(1,id);
18.         statement.setString(1,name);
19.     }
20.     @Override
21.     public void readFields(ResultSet resultSet) throws SQLException {
22.         id = resultSet.getInt("id");
23.         name = resultSet.getString("name");
24.     }
    //此处省去了对 id,name 的 getter、setter
25. }
```

编写 Mapper，此类只是接收到数据以后直接输出即可。

【代码 5-6】DBMapper.java

```
1.  package org.hadoop.input.db;
2.  public class DBMapper extends Mapper<LongWritable,StudWritable, Text, NullWritable> {
3.      private Text outKey = new Text();
4.      @Override
5.      protected void map(LongWritable key, StudWritable value, Context context)
6.              throws IOException, InterruptedException {
7.          outKey.set(key.get()+"|"+value.getId()+"|"+value.getName());
8.          context.write(outKey,NullWritable.get());
9.      }
10. }
```

编写 Reducer，同样也是接收到数据以后直接输出即可。

【代码 5-7】DBReducer.java

```
1.  package org.hadoop.input.db;
2.  public class DBReducer extends Reducer<Text, NullWritable, Text, NullWritable> {
3.      @Override
4.      protected void reduce(Text key, Iterable<NullWritable> values, Context context)
5.              throws IOException, InterruptedException {
6.          context.write(key,NullWritable.get());
7.      }
8.  }
```

编写 MapRecucer 程序。

【代码 5-8】DBMR.java

```
1.  package org.hadoop.input.db;
2.  public class DBMR extends Configured implements Tool {
3.      @Override
4.      public int run(String[] args) throws Exception {
5.          Configuration conf = getConf();
6.          String url = "jdbc:mysql://127.0.0.1:3306/hadoop?characterEncoding=UTF-8&serverTimezone=Asia/Shanghai";
7.          DBConfiguration.configureDB(conf, "com.mysql.cj.jdbc.Driver", url, "root", "123456");
8.          Path dest = new Path("D:/a/out003");
9.          FileSystem fs = FileSystem.get(conf);
10.         if (fs.exists(dest)) {
11.             fs.delete(dest, true);
12.         }
13.         Job job = Job.getInstance(conf, "db");
14.         job.setJarByClass(getClass());
15.         job.setMapperClass(DBMapper.class);
16.         job.setMapOutputKeyClass(Text.class);
17.         job.setMapOutputValueClass(NullWritable.class);
18.         job.setReducerClass(DBReducer.class);
19.         job.setOutputKeyClass(Text.class);
20.         job.setOutputValueClass(NullWritable.class);
21.         job.setInputFormatClass(DBInputFormat.class);
22.         job.setOutputFormatClass(TextOutputFormat.class);
23.         DBInputFormat.setInput(job, StudWritable.class, "stud", null, "id", "id", "name");
24.         FileOutputFormat.setOutputPath(job, dest);
25.         int code = job.waitForCompletion(true) ? 0 : 1;
26.         return code;
27.     }
28.     public static void main(String[] args) throws Exception {
29.         int run = ToolRunner.run(new DBMR(), args);
30.         System.exit(run);
31.     }
32. }
```

现在执行 DBMR.java 程序，可以在输出的目录下查看到如下输出内容：

1. 0|1|Jack
2. 1|2|Mary

5.1.4 自定义输出流

自定义输出流需要继承 OutputFormat。以下自定义一个输出流，直接将数据输出到指定的文件中。

输出对象的 Key、Value 应该为 Reducer 结果的输出 Key、Value 类型。

【代码 5-9】MyOutputFormat.java

```
1.  package org.hadoop.out;
2.  public class MyOutputFormat extends FileOutputFormat<LongWritable, Text> {
3.      @Override
4.      public RecordWriter<LongWritable, Text> getRecordWriter(TaskAttemptContext job) throws IOException, InterruptedException {
5.          FileSystem fs = FileSystem.get(job.getConfiguration());
6.          Path path = getOutputPath(job);//获取输出目录
7.          path = new Path(path, "a.txt");//任意定义一个输出的文件
8.          PrintWriter writer = new PrintWriter(fs.create(path), true);
9.          MyRecordWriter myRecordWriter = new MyRecordWriter(writer);
10.         return myRecordWriter;
11.     }
12.     public static class MyRecordWriter extends RecordWriter<LongWritable, Text> {
13.         PrintWriter printWriter = null;
14.
15.         public MyRecordWriter(PrintWriter printWriter) {
16.             this.printWriter = printWriter;
17.         }
18.         @Override
19.         public void write(LongWritable key, Text value) throws IOException, InterruptedException {
20.             //仅在输出数据添加追加字符串来演示自定义输出流
21.             printWriter.println(">>" + key.toString() + "\t" + value);
22.         }
23.         @Override
24.         public void close(TaskAttemptContext context) throws IOException, InterruptedException {
25.             if (printWriter != null) {
26.                 printWriter.close();
27.             }
28.         }
29.     }
30. }
```

然后在 MapReduce 中使用这个输出流对象，为了简单起见，我们直接使用默认的 Mapper 和

Reducer，此时默认输出的 Key 和 Value 为 LongWritable 和 Text。

【代码 5-10】MyOutputMR.java

```java
1.  package org.hadoop.out;
2.  public class MyOutputMR extends Configured implements Tool {
3.      @Override
4.      public int run(String[] args) throws Exception {
5.          if(args.length<2){
6.              System.out.println("用法：输入 输出");
7.              ToolRunner.printGenericCommandUsage(System.out);
8.              return -1;
9.          }
10.         Path dest = new Path(args[1]);
11.         Configuration conf = getConf();
12.         FileSystem fs = FileSystem.get(conf);
13.         if(fs.exists(dest)){
14.             fs.delete(dest,true);
15.         }
16.         Job job = Job.getInstance(conf,"MyOutput");
17.         job.setJarByClass(this.getClass());
18.         job.setOutputKeyClass(LongWritable.class);
19.         job.setOutputValueClass(Text.class);
20.         job.setInputFormatClass(TextInputFormat.class);
21.         //设置为自定义输出流对象
22.         job.setOutputFormatClass(MyOutputFormat.class);
23.         FileInputFormat.setInputPaths(job,new Path(args[0]));
24.         FileOutputFormat.setOutputPath(job,new Path(args[1]));
25.         return job.waitForCompletion(true)?0:1;
26.     }
27.     public static void main(String[] args) throws Exception {
28.         int r = ToolRunner.run(new MyOutputMR(),args);
29.         System.exit(r);
30.     }
31. }
```

运行完成以后，查看输出的结果，如果前面都带有>>，则表示输出成功。

5.2 顺序文件 SequenceFile 的读写

顺序文件是以 key-value 形式存在的文件。Key 和 Value 都必须是 Writable 的子类。sequenceFile 文件是 Hadoop 用来存储二进制形式的键值对而设计的一种平面文件（Flat File）。可以把 SequenceFile 当作是一个容器，把所有的文件打包到 SequenceFile 类中可以高效地对小文件进行存储和处理。SequenceFile 文件并不按照其存储的 Key 进行排序存储，SequenceFile 的内部类 Writer 提供了 append 功能。SequenceFile 中的 Key 和 Value 可以是任意类型 Writable 或者是自定义 Writable。在存储结构上，SequenceFile 主要由一个 Header 后跟多条 Record 组成。Header 主要包含了 Key classname、value classname、存储压缩算法、用户自定义元数据等信息，此外，还包含了一些同步标识，用于快速定位到记录的边界。每条 Record 以键值对的方式进行存储，用来表示它的字符数组可以一次解析成：记录的长度、Key 的长度、Key 值和 Value 值，并且 Value 值的结构取决于该记录是否被压缩。

SequenceFile 的优点如下：

- 支持基于记录（Record）或块（Block）的数据压缩。
- 支持 splitable，能够作为 MapReduce 的输入分片。
- 修改简单，主要负责修改相应的业务逻辑，而不用考虑具体的存储格式。

SequenceFile 的缺点如下：

- 需要一个合并文件的过程，且合并后的文件不方便查看。

5.2.1 生成一个顺序文件

可以借助 Hadoop 的 SequenceFile 类，快速创建一个顺序文件。以下代码生成一个顺序文件，并写入几行数据。

【代码 5-11】CreateSequenceFile.java

```
1.  package org.hadoop.seq;
2.  public class CreateSequenceFile {
3.      public static void main(String[] args) throws Exception {
4.          Configuration config = new Configuration();
5.          //设置文件名和 Key、Value 类型
6.          Writer.Option op1 = Writer.file(new Path("file:///D:/a/1.seq"));
7.          Writer.Option op2 = Writer.keyClass(IntWritable.class);
8.          Writer.Option op3 = Writer.valueClass(Text.class);
9.          SequenceFile.Writer w = SequenceFile.createWriter(config, op1, op2, op3);
10.         w.append(new IntWritable(1),
11.                 new org.apache.hadoop.io.Text("Hello"));
12.         w.append(new IntWritable(20),
13.                 new org.apache.hadoop.io.Text("Second Row"));
14.         w.append(new IntWritable(3),
15.                 new org.apache.hadoop.io.Text("last Row"));
16.         w.close();
17.     }
18. }
```

查看顺序文件，使用 -text 命令：

```
D:\a>hdfs dfs -text file:///D:/a/1.seq
1       Hello
20      Second Row
3       last Row
```

5.2.2 读取顺序文件

同样地，可以借助 SequenceFile 类读取顺序文件。

【代码 5-12】ReadSequenceFile.java

```
1.  package org.hadoop.seq;
2.  public class ReadSequenceFile {
3.      public static void main(String[] args) throws Exception {
4.          Configuration config = new Configuration();
5.          Reader.Option op1 = Reader.file(new Path("file:///D:/a/1.seq"));
6.          SequenceFile.Reader r = new SequenceFile.Reader(config,op1);
7.          IntWritable key = new IntWritable();
8.          Text value = new Text();
9.          while(r.next(key, value)) {
10.             System.out.println(key+","+value);
11.         }
12.         r.close();
13.     }
14. }
```

执行代码后,查看输出即为前面写入的数据内容。

5.2.3 获取 Key/Value 类型

如果已经知道顺序文件的 Key 和 Value 类型,则可以直接使用 Reader.next(key,value)来读取这些数据;如果不知道,可以通过下面代码获取:

```
reader.getKeyClass() 获取 key 类型
reader.getValueClass() 获取 value 类型
```

【代码 5-13】ReadSequenceFIle2.java

```
1.  package org.hadoop.seq;
2.  /**
3.   * 通过 reader.getValueClass()<br>
4.   * getKeyClass()获取类型
5.   */
6.  public class ReadSequenceFile2 {
7.      public static void main(String[] args)  throws Exception{
8.          Configuration config = new Configuration();
9.          Reader.Option op1 = Reader.file(new Path("file:///D:/a/1.seq"));
10.         SequenceFile.Reader reader = new SequenceFile.Reader(config,op1);
11.         //获取 Key 和 Value 类型
12.         Writable key =  (Writable) reader.getKeyClass().newInstance();
13.         Writable value = (Writable)reader.getValueClass().newInstance();
14.         //读取数据
15.         while(reader.next(key, value)) {
16.             System.out.println(key.toString()+"\t"+value.toString());
17.         }
18.         reader.close();
19.     }
20. }
```

5.2.4 使用 SequenceFileInputFormat 读取数据

Hadoop 提供了一个 SequenceFileInputFormat 类，可以用来读取 SequenceFile。此类是 InputFormat 的子类。它们的继承关系如图 5-2 所示。

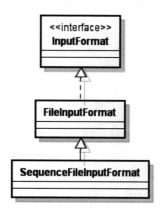

图 5-2

现在我们在 MapReduce 程序中使用 SequenceFileInputFormat。此时开发的 Mapper 程序接收的 key 应该为 SequenceFile 文件的 Key 和 Value 类型。以下代码读取 SequenceFile。

【代码 5-14】ReadSequenceFileMR.java

```
1.  package org.hadoop.seq;
2.  public class ReadSequenceFileMR extends Configured implements Tool {
3.      @Override
4.      public int run(String[] args) throws Exception {
5.          Configuration config = getConf();
6.          Path dest = new Path("D:/a/out001");
7.          FileSystem fs = FileSystem.get(config);
8.          if (fs.exists(dest)) {
9.              fs.delete(dest, true);
10.         }
11.         //声明 Job
12.         Job job = Job.getInstance(config, "Seq");
13.         job.setJarByClass(getClass());
14.         //设置 Mapper
15.         job.setMapperClass(Demo10Mapper.class);
16.         job.setMapOutputKeyClass(IntWritable.class);
17.         job.setMapOutputValueClass(Text.class);
18.         //设置 Reducer
19.         job.setReducerClass(Demo10Reducer.class);
20.         job.setOutputKeyClass(IntWritable.class);
21.         job.setOutputValueClass(Text.class);
22.         //设置输入流为 SequenceFile
23.         job.setInputFormatClass(SequenceFileInputFormat.class);
24.         job.setOutputFormatClass(TextOutputFormat.class);
25.         //设置输入/输出目录
26.         FileInputFormat.setInputPaths(job, new Path("D:/a/1.seq"));
27.         FileOutputFormat.setOutputPath(job, dest);
28.         int code = job.waitForCompletion(true) ? 0 : 1;
```

```
29.         return code;
30.     }
31.     public static class Demo10Mapper
32.         extends Mapper<IntWritable, Text, IntWritable, Text> {
33.         @Override
34.         public void map(IntWritable key, Text value, Mapper<IntWritable, Text, IntWritable, Text>.Context context)
35.             throws IOException, InterruptedException {
36.             context.write(key, value);
37.         }
38.     }
39.     public static class Demo10Reducer
40.         extends Reducer<IntWritable, Text, IntWritable, Text> {
41.         @Override
42.         protected void reduce(IntWritable key, Iterable<Text> values,
43.                 Reducer<IntWritable, Text, IntWritable, Text>.Context context)
44.             throws IOException, InterruptedException {
45.             context.write(key, values.iterator().next());
46.         }
47.     }
48.     public static void main(String[] args) throws Exception {
49.         int code = ToolRunner.run(new ReadSequenceFileMR(), args);
50.         System.exit(code);
51.     }
52. }
```

SequenceFileOutputFormat 可用于输出顺序文件。继承关系如下：

```
OutputFormat
FileOutputFormat
SequenceFileOutputFormat
```

以下示例将文本文件转存成顺序文件，以下完整代码中使用了默认了 Mapper 和默认的 Reducer，只是在 Job 中设置了 job.setOutputFormatClass(SequenceFileOutputFormat.class)。

【代码 5-15】SequeceFileOutMR.java

```
1.  package org.hadoop.seq;
2.  public class SequenceFileOutMR extends Configured implements Tool {
3.      @Override
4.      public int run(String[] args) throws Exception {
5.          Configuration config = getConf();
6.          FileSystem fs = FileSystem.get(config);
7.          Path dest  = new Path("D:/a/out002");
8.          if(fs.exists(dest)) {
9.              fs.delete(dest, true);
10.         }
11.         Job job = Job.getInstance(config,"Seq");
12.         job.setJarByClass(getClass());
13.         job.setInputFormatClass(TextInputFormat.class);
14.         job.setOutputFormatClass(SequenceFileOutputFormat.class);
15.         FileInputFormat.setInputPaths(job, new Path("D:/a/a.txt"));
16.         FileOutputFormat.setOutputPath(job, dest);
17.         int code = job.waitForCompletion(true)?0:1;
```

```
18.            return code;
19.        }
20.    public static void main(String[] args) throws Exception {
21.        int code = ToolRunner.run(new SequenceFileOutMR(), args);
22.        System.exit(code);
23.    }
24. }
```

执行完成以后，通过 **hdfs dfs -text** 查看输出的内容：

```
D:\a\out002>hdfs dfs -text part-r-00000
0       Hello This Is First Line .
28      This Example show how
51      to implements Writable
```

5.3 本章小结

到目前为止，我们已经学会了运行一个单一节点的 Hadoop。表 5-1 所示是 Hadoop 在伪分布式上的一些配置，供读者参考。

表 5-1　Hadoop 在伪分布式上的一些配置

软件或配置	说　明
修改主机名称	hostnamectl
配置本地 DNS	/etc/hosts 192.168.56.201　　　server201
禁止使用防火墙	firewall-cmd --state systemctl stop firewalld.service systemctl disable firewalld.service
SSH 免密码登录 ~/.ssh/ id_rsa　id_rsa.pub authorized_keys	ssh-keygen -t rsa Ssh-copy-id server201
格式化 Hadoop 的文件系统	hdfs namenode -format 或 hadoop namenode -format
开发 MapReduce	extends Mapper<Key1,Value1,Key2,Value2> extends Reducer<Key3,Value3,Key4,Value4> 然后使用 job 来添加 Mapper 和 Reducer
开发 Partitioner 分区管理	extends Partitioner<Key2,Value2>
开发 Combiner	Combiner 就是一个 Reducer，用于在 Map 和 Reduce 之前进行数据合并
自定义 InputFormat	extends FileInputFormat<Key1,Value1> 注册自定义输入类
自定义序列化类	extends Writable，只是序列化 extends WritableComparable，排序
查看 HDFS 保存的文件信息	$hadoop_tmp_dir/dfs/data 目录下，128MB 为一个文件块大小

第 6 章

Hadoop 分布式集群配置

本章主要内容：

- Hadoop 分布式集群配置。

6.1　Hadoop 集群

在 Hadoop 的集群中，有一个 NameNode，一个 ResourceManager。在高可靠的集群环境中，可以拥有两个 NameNode 和两个 ResourceManager；在 Hadoop 3 版本以后，同一个 NameService 可以拥有 3 个 NameNode，这将在后面的章节中讲解。由于 NameNode 和 ResourceManager 是两个主要的服务，建议将它们部署到不同的服务器上。

非高可靠集群可用于快速学习 Hadoop 的集群方式，在深入了解 Hadoop 集群运行的基本原理上非常有用。以下将分步骤详解如何搭建 Hadoop 的非高可靠集群。

步骤 01 准备工作。

以三台服务器为集群环境，做以下准备工作。

（1）所有主机安装 JDK1.8+。建议将 JDK 安装到不同的主机的相同目录下，这样可以减少修改配置文件的次数。

（2）在主节点（即执行 start-dfs.sh 和 start-yarn.sh 的主机）上向所有其他主机做 SSH 免密码登录。

（3）修改所有主机的主机名称。

（4）配置所有主机的 hosts 文件，添加主机名和 IP 的映射如下：

```
192.168.56.101 server101
192.168.56.102 server102
192.168.56.103 server103
```

（5）关闭所有主机上的防火墙，使用以下命令：

```
systemctl stop firewalld
systemctl disable firewalld
```

（6）完整的配置如表 6-1 所示。

表 6-1 集群主机配置表

IP/主机名	虚拟机	进程	软件
192.168.56.101/server101	8GB 内存，2 核	NameNode SecondaryNameNode ResourceManager DataNode NodeManager	JDK HADOOP
192.168.56.102/server102	2GB 内存，1 核	DataNode NodeManager	JDK HADOOP
192.168.56.103/server103	2GB 内存，1 核	DataNode NodeManager	JDK Hadoop

从表 6-1 可以看出，sserver101 运行的进程比较多，且 NameNode 运行在上面，所以这台主机需要更多的内存。

特别说明，第 7 章的 ZooKeeper 集群安装，第 9 章 HBase 的集群安装以及第 11 章 Spark 集群安装与本章集群主机配置表相同，即都采用表 6-1 所示的三台主机，具体配置内容详见各章节具体描述。

步骤 02 在 server101 上安装 Hadoop。

可以将 Hadoop 安装到任意的目录下，如在根目录下创建/app 然后授予 hadoop 用户即可。
将 hadoop-3.2.2.tar.gz 解压到/app 目录下，并配置/app 目录属于 hadoop 用户。

```
$ sudo tar -zxvf hadoop3.2.2.tag.gz -C /app/
```

将/app 目录及子目录，授权给 hadoop 用户和 hadoop 组：

```
$suto chown hadoop:hadoop -R /app
```

配置 hadoop-env.sh 文件：

```
export JAVA_HOME=/usr/java/jdk1.8.0_281
```

配置文件 core-site.xml：

```
<configuration>
    <property>
        <name>fs.defaultFS</name>
        <value>hdfs://server101:8020</value>
    </property>
    <property>
        <name>hadoop.tmp.dir</name>
        <value>/app/datas/hadoop</value>
    </property>
```

```
</configuration>
```

配置文件 hdfs-site.xml：

```
<configuration>
    <property>
        <name>dfs.replication</name>
        <value>3</value>
    </property>
    <property>
        <name>dfs.permissions.enabled</name>
        <value>false</value>
    </property>
</configuration>
```

配置文件 mapred-site.xml：

```
<configuration>
    <property>
        <name>mapreduce.framework.name</name>
        <value>yarn</value>
    </property>
</configuration>
```

配置文件 yarn-site.xml：

```
<configuration>
    <property>
        <name>yarn.nodemanager.aux-services</name>
        <value>mapreduce_shuffle</value>
    </property>
    <property>
        <name>yarn.resourcemanager.hostname</name>
        <value>server101</value>
    </property>
    <property>
        <name>yarn.application.classpath</name>
        <value>请自行执行 hadoop classpath 命令并将结果填入</value>
    </property>
</configuration>
```

配置 workers 配置文件：workers 配置文件用于配置执行 DataNode 和 NodeManager 的节点。

```
server101
server102
server103
```

步骤 03 使用 scp 将 Hadoop 分发到其他主机。

由于 scp 会在网络上传递文件，而 hadoop/share/doc 目录下都是文档，没有必要进行复制，所以可以删除这个目录。

删除 doc 目录：

```
$ rm -rf /app/hadoop-3.2.2/share/doc
```

然后复制 server101 的文件到其他两台主机的相同目录下：

```
$scp -r /app/hadoop-3.2.2   server102:/app/
$scp -r /app/hadoop-3.2.2   server103:/app/
```

步骤 04 在 server101 上格式化 NameNode。

首先需要在 server101 上配置 Hadoop 的环境变量。

```
$ sudo vim /etc/profile
```

在文件最后追加：

```
export HADOOP_HOME=/app/hadoop-3.2.2
export PATH=$PATH:$HADOOP_HOME/bin
```

在 server101 上执行 namenode 初始化命令：

```
$ hdfs namenode -format
```

步骤 05 启动 HDFS 和 YARN。

在 server101 上执行启动工作，由于配置了集群，此启动过程会以 SSH 方式登录其他两台主机，并分别启动 DataNode 和 NodeManager。

```
$ /app/hadoop-3.2.2/sbin/start-dfs.sh
$ /app/hadoop-3.2.2/sbin/start-yarn.sh
```

启动完成后，通过宿主机的浏览器，查看 9870 端口，会显示集群情况，如图 6-1 所示。
访问地址 http://server101:9870/，会发现三个 DataNode 节点同时存在。

图 6-1

查看 8088 端口，会发现三个 NodeManager 同时存在，如图 6-2 所示。

```
http://192.168.1.10:8088
```

图 6-2

最后，建议执行 MapReduce 测试一下集群，比如执行以下 wordcount 示例，如果可以顺序执行完成，则说明整个集群的配置都是正确的。

```
$ yarn jar hadoop-mapreduce-examples-3.2.2.jar wordcount /test/ /out002
```

6.2 本章小结

本章主要讲解了以下内容：

- Hadoop 集群需要先安装 JDK，并配置 JAVA 环境变量。
- 配置过程共修改 Hadoop 的 6 个配置文件，此过程与 Hadoop 的伪分布式相同，只有 works 配置文件中添加三个主机名称。

这种集群的问题是：一旦 NameNode 宕机，整个集群即不能运行；一旦 ResourceManager 宕机，整个 Mapreduce 将不能运行。这就需要搭建 HA 高可靠集群。

第 7 章

Hadoop 高可用集群搭建

本章主要内容：

- ZooKeeper 简介。
- ZooKeeper 分布式安装。
- 使用 ZooKeeper 实现 Hadoop 的高可靠集群。

7.1 ZooKeeper 简介

ZooKeeper 是一个分布式的、开放源码的分布式应用程序协调服务，它包含一个简单的原语集，分布式应用程序可以基于它实现同步服务，配置维护和命名服务等。ZooKeeper 是 Hadoop 的一个子项目。在分布式应用中，由于工程师不能很好地使用锁机制，以及基于消息的协调机制不适合在某些应用中使用，因此需要有一种可靠的、可扩展的、分布式的、可配置的协调机制来统一系统的状态。ZooKeeper 的目的就在于此。

目前，在分布式协调技术方面做得比较好的就是 Google 的 Chubby，还有 Apache 的 ZooKeeper，它们都是分布式锁的实现者，如图 7-1 所示。有人会问，既然有了 Chubby 为什么还要弄一个 ZooKeeper，难道 Chubby 做得不够好吗？不是这样的，主要是 Chubby 是非开源的，Google 自家使用。后来雅虎模仿 Chubby 开发出了 ZooKeeper，也实现了类似的分布式锁的功能，并且将 ZooKeeper 作为一种开源的程序捐献给了 Apache，那么这样就可以使用 ZooKeeper 所提供锁服务，而且它在分布式领域久经考验，其可靠性、可用性都是经过理论和实践验证过的。所以在构建一些分布式系统的时候，就可以以这类系统为起点来构建我们的系统，这将节省不少成本，而且 Bug 也会更少。

本章学习 ZooKeeper 的主要目的就是为了搭建 Hadoop 的高可靠集群。

图 7-1

ZooKeeper 中的角色主要有三类，如表 7-1 所示。

表 7-1 ZooKeeper 中的角色

角色		说明
领导者（Leader）		负责进行投票的发起和决议，更新系统状态
学习者（Learner）	跟随者（Follower）	Follower 用于接收客户请求并向客户返回结果，且在选主过程中，参考投票
	观察者（ObServer）	ObServer 可以接收客户端连接，将写请求转发给 Leader 节点。但 ObServer 不参与投票过程，只同步 Leader 的状态。ObServer 的目的是为了扩展系统，提高读取速度
客户端（Client）		请求发起方

系统模型如图 7-2 所示。

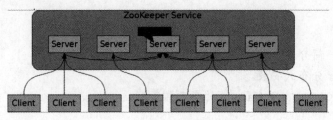

图 7-2

1. ZooKeeper 的时间

在 ZooKeeper 中，有多种记录时间的形式，其中包含以下几个主要属性：

（1）Zxid

导致 ZooKeeper 节点状态改变的每一个操作都将使节点接收到一个 Zxid 格式的时间戳，并且这个时间戳全局有序。也就是说，每个对节点的改变都将产生一个唯一的 Zxid。如果 Zxid1 的值小于 Zxid2 的值，那么 Zxid1 所对应的事件发生在 Zxid2 所对应的事件之前。实际上，ZooKeeper 的每个节点维护者三个 Zxid 值分别为：cZxid、mZxid、pZxid。

① cZxid：是节点的创建时间所对应的 Zxid 格式时间戳。
② mZxid：是节点的修改时间所对应的 Zxid 格式时间戳。

实现中 Zxid 是一个 64 位的数字，它的高 32 位是 epoch 用来标识 Leader 关系是否改变，每次一个 Leader 被选出来，它都会有一个新的 epoch。低 32 位是个递增计数。

（2）版本号

对节点的每一个操作都将导致这个节点的版本号增加。每个节点维护着三个版本号，它们分别为：

① version：节点数据版本号。
② cversion：子节点版本号。
③ aversion：节点所拥有的 ACL 版本号。

2. ZooKeeper 节点属性

一个节点自身拥有表示其状态的许多重要属性，如表 7-2 所示。

表 7-2　ZooKeeper 节点属性

属　性	说　明
czxid	节点被创建的 zxid
mzxid	节点被修改的 zxid
ctime	节点被创建的时间
mtime	节点被修改的时间
version	节点被修改的版本号
cversion	节点所拥有的子节点被修改的版本号
aversion	节点的 ACL 被修改的版本号
ephemeralOwner	如果此节点为临时节点，那么它的值为这个节点拥有者的会话 ID，否则值为 0
dataLength	节点数的长度
numChildren	节点拥有的子节点的长度
pzxid	最新修改的 zxid

3. ZooKeeper 服务中操作

在 ZooKeeper 中有 9 个基本操作，如表 7-3 所示。

表 7-3　ZooKeeper 中 9 个基本操作

操　作	说　明
Create	创建 znode（父 znode 必须存在）
Delete	删除 znode（znode 没有子节点）
Exists	判断 znode 是否存在
getACL/setACL	读取或设置 ACL
getChildren	获取 znode 所有子节点的列表
getData/setData	获取/设置 znode 相关数据
Sync	客户端与 ZooKeeper 同步

4. ZooKeeper 的配置文件

ZooKeeper 的默认配置文件为$ZOOKEEPER_HOME/conf/zoo.cfg。其中各配置项的含义，解释如下：

（1）tickTime：CS 通信心跳时间

ZooKeeper 服务器之间或客户端与服务器之间维持心跳的时间间隔，也就是每个 tickTime 时间就会发送一个心跳。tickTime 以毫秒为单位，如 tickTime=2000。

（2）initLimit：LF 初始通信时限

集群中的 follower 服务器（F）与 leader 服务器（L）之间初始连接时能容忍的最多心跳数（tickTime 的数量），如 initLimit=5。

（3）syncLimit：LF 同步通信时限

集群中的 Follower 服务器与 Leader 服务器之间请求和应答之间能容忍的最多心跳数（tickTime 的数量），如 syncLimit=2。

（4）dataDir：数据文件目录

ZooKeeper 保存数据的目录。默认情况下，ZooKeeper 将写数据的日志文件也保存在这个目录里，如设置 dataDir=/home/zoo/SomeData。

（5）clientPort：客户端连接端口

客户端连接 ZooKeeper 服务器的端口，ZooKeeper 会监听这个端口，接受客户端的访问请求，如 clientPort=2181。

（6）服务器名称与地址：集群信息

集群信息包括服务器编号、服务器地址、LF 通信端口、选举端口。这个配置项的书写格式比较特殊，规则如下：

```
server.N=YYY:A:B
```

如配置以下示例：

```
server.1=ip:2888:3888
server.2=ip:2888:3888
server.3=ip:2888:3888
```

7.2　ZooKeeper 集群安装

想要安装 ZooKeeper 集群，首先就要在多台机器上都安装 ZooKeeper。为了便于记忆，可以将 ZooKeeper 安装到相同的目录下，如/app 目录下。

在每一个 ZooKeeper 的 dataDir 目录下，创建一个 myid 文件，里面保存的是当前 ZooKeeper 节点的 ID。ID 不一定从 1 开始。本示例中的 ID 选取 IP 地址最后一段作为 ID，以便于记忆。

步骤 01 配置 zoo.cfg。

配置三台 CentOS7 主机，三台主机的配置与第 6 章分布式集群的三台主机配置相同，每台主机都需要关闭防火墙。然后修改$ZOOKEEPER_HOME/conf/zoo.cfg，在文件最后追加以下配置：

```
#配置 ZooKeeper 数据保存目录
dataDir=/app/datas/zk
```

配置 ZooKeeper 集群：

```
server.101=192.168.56.101:2888:3888
server.102=192.168.56.102:2888:3888
server.103=192.168.56.103:2888:3888
```

步骤 02 使用 scp 将文件发送到其他两台机器。

```
$ scp -r zookeeper-3.6.2  server102:/app/
$ scp -r zookeeper-3.6.2  server103:/app/
```

然后修改每一个 dataDir 目录下的 myid 文件。在 server101 主机上的 myid 中添加 101（笔者是为了方便记忆才取此 ID 的，此 ID 等于 IP 的地址，建议读者可以从 1 开始），即：

```
echo 101 > /app/datas/zk/myid
```

以此类推。

步骤 03 现在分别启动三台主机的 ZooKeeper。

启动命令：

```
$ ./zkServer.sh start
```

然后查看状态，使用 status 检查状态：

```
[hadoop@server101 bin]$ ./zkServer.sh status
Mode: leader    #这是 leader 其他两台为 follower
```

现在就可以同步测试了。在一台虚拟机上进行操作，查看其他两台的同步情况。

步骤 04 测试操作。

登录客户端：

```
$ zkCli.sh
```

显示当前所有目录：

```
[hadoop: localhost 2181] ls /
[zookeeper]
```

创建一个新的目录，且写入数据：

```
[hadoop: localhost : 2181 ] create /test TestData
```

再次显示当前根目录下的所有数据。登录其他主机，如果查看到相同的结果，即表示已经同步。

```
[hadoop: localhost : 2181 ] ls /
[test, zookeeper]
```

ZooKeeper 的命令很多，可以使用 help 查看所有可使用的命令。

```
[zhadoopk: localhost : 2181] help
```

通过上面的配置可以看出，ZooKeeper 的集群配置相对比较简单。只是配置 zoo.cfg 并指定所有服务节点即可。然后在每一个节点的 data 目录下，将当前 id 写入到 myid 文件中即可。

7.3 znode 节点类型

znode 的节点类型分为持久节点、顺序节点和临时节点。默认创建的节点都是持久节点，使用 -s 参数可以创建一个顺序节点，使用 -e 参数可以创建临时节点。

创建顺序节点：

```
[hadoop: localhost:2181(CONNECTED) 13] create -s /t ""
Created /t0000000004
```

创建临时节点,客户端退出时自动被删除:

```
[hadoop: localhost:2181(CONNECTED) 2] create -e /tt ""
Created /tt
```

7.4 观察节点

观察者 ObServer 可以观察到节点的变化,常用命令如下:

```
stat path [watch]
ls path [watch]
ls2 path [watch]
get path [watch]
```

上面的命令后面都有一个[watch],即可以在命令行观察节点的变化。

首先通过某个设置,在后面直接添加 watch:

```
[hadoop: localhost:2181(CONNECTED) 23] ls /one watch
[]
```

然后在另一个的客户端,添加一个子节点:

```
[hadoop: localhost:2181(CONNECTED) 4] create /one/1 Some
Created /one/1
```

现在可以观察到节点显示的数据:

```
[hadoop: localhost:2181(CONNECTED) 24]
WATCHER::
WatchedEvent state:SyncConnected type:NodeChildrenChanged path:/one
```

7.5 配置 Hadoop 高可靠集群

Hadoop 2.0 增加了很多特性,比如 HDFS HA、YARN 都可以配置 HA（高可靠）。在 Hadoop 2.x 版本时一个 NameNodeService 只能有两个 NameNode,而到了 3.0 版本一个 NameService 可以有三个 NameNode 或是更多。以下是官方网站说明:

```
The minimum number of NameNodes for HA is two, but you can configure more. Its
suggested to not exceed 5 - with a recommended 3 NameNodes - due to communication
overheads.
```

配置 Hadoop 高可靠将会启动一些新的进程,具体如下:

- ZKFC（DFSZKFailoverController）：Hadoop 进程,ZooKeeper 通过与 ZKFC 通信,获取 NameNode 的活动状态。如果 ZKFC 认为 NameNode 已经宕机,将会通过 ZooKeeper 开启新的选举程序,选择出新的主 NameNode。
- QJM（Quorum Journal Manager）：用于同步主 NameNode 的日志数据保存到备份

NameNode 中。在高可靠情况下，将不再使用 SecondaryNameNode 程序，多个 NameNode 之间互相备份。QJM 负责它们之间的通信数据。

由于配置 Hadoop 的高可靠集群比较复杂，所以我们需要做好配置前的规划。
配置过程列表，以下是具体的配置规划：

- 配置从一台主机到其他主机的 SSH 免密码登录。主要是执行 start-dfs.sh 和 start-yarn.sh 的主机到其他主机的 SSH 免密码登录。
- 关闭所有主机的防火墙，仅非生产环境。
- 配置所有主机的静态地址和 hosts 文件。
- 所有主机上安装 JDK1.8+，并配置 Java 环境变量。
- 至少在三台主机上安装好 ZooKeeper 并启动 ZooKeeper 集群。
- 在一台主机上配置好 Hadoop 的所有配置文件，并通过 scp 分发到所有主机的相同目录下。
- 所有主机配置 Hadoop 环境变量。
- 启动 JournalNode。
- 在某台配置了 NameNode 的主机上格式化 NameNode。
- 然后将格式化后的目录复制到其他的主机（注意只是配置了 NameNode 的主机）。
- 格式化 ZKFC。
- 启动 HDFS，启动 YARN。

步骤 01 配置计划表：

IP/主机名	软件	进程
192.168.56.101 Server101	JDK1.8+ Zookeeper3.6.2 Hadoop3.2.2	QuorumPeerMain NameNode ZKFC QJM ResourceManager NodeManager DataNode
192.168.56.102 Server102	JDK1.8+ Zookeeper3.6.2 Hadoop3.2.2	QuorumPeerMain NameNode ZKFC QJM ResourceManager NodeManager DataNode
192.168.56.103 Server103	JDK1.8+ Zookeeper3.6.2 Hadoop3.2.2	QuourmPeerMain NameNode ZKFC QJM ResourceManager NodeManager DataNode

步骤 02 前期准备。

所有主机关闭防火墙：

```
$ sudo sytemctl stop firewalld
$ sudo systemctl disable firewalld
```

所有主机安装 JDK1.8+，并配置环境变量：

```
$ sudo tar -zxvf ~/jdk1.8-281.tag.gz -C /usr/java/
$ sudo vim /etc/provfile
export JAVA_HOHE=/usr/java/jdk1.8-281
exoort PATH=$PATH:$JAVA_HOME/bin
```

所有主机设置静态 IP 地址，修改主机名称，需要分别修改三台主机。下面以一台主机为例：

```
$ sudo vim /etc/sysconfig/network-scripts/ifcfg-enp0s8
IPADDR=192.168.56.101
$ sudo systemctl hostnamectl set-hostname server101
$ sudo vim /etc/hosts
192.168.56.101 server101
192.168.56.102 server102
192.168.56.103 server103
```

设置所有主机 selinux=disabled：

```
$ sudo vim /etc/selinux/config
selinux=disabled
```

安装好 ZooKeeper 集群，并启动。请参考 7.3 节。

步骤 03 配置 hadoop-env.sh 文件。

在 hadoop-env.sh 文件中，添加 JAVA_HOME 环境变量：

```
export JAVA_HOME=/usr/local/java/jdk1.8.0_281
```

步骤 04 配置 core-site.xml 文件。

与之前的配置一样，配置 Hadoop 的 core-site.xml 文件，但请注意，之前配置的 fs.defaultFS 的值为 hdfs://server101:8020，在集群模式下，应该配置为 hdfs://cluster。其中 cluster 为任意设置的名称，请牢记此名称，后台将会使用此名称来配置 NameService。

```
1.  <configuration>
2.      <property>
3.          <name>fs.defaultFS</name>
4.          <value>hdfs://cluster</value>
5.      </property>
6.      <property>
7.          <name>hadoop.tmp.dir</name>
8.          <value>/app/datas/hadoop</value>
9.      </property>
10.     <property>
11.         <name>ha.zookeeper.quorum</name>
12.         <value>server101:2181,server102:2181,server103:2181</value>
13.     </property>
```

```
14.    </configuration>
```

配置说明：

- 第 4 行代码：指定集群名称，其中 cluster 可以是任意的名称，但请牢记此名称，后面将会用到。
- 第 8 行代码：与之前一样，配置 NameNode 保存数据的本地目录，此目录使用 hdfs namenode -format 命令格式化后，将会有数据。将格式化完成的目录，通过 scp 复制到其他 NameNode 节点的相同目录下，这样做的目的是为了保证 Hadoop 集群 ID 的唯一。
- 第 12 行代码：指定 ZooKeeper 的集群地址。由于我们主机不是很多，所以这里 ZooKeeper 的地址很多与 Hadoop 存储计算结果相同，在正式的生产环境下，此 ZooKeeper 节点应该有独立的主机。

步骤 05 配置 hdfs-site.xml 文件。

这里的配置信息比较多，请注意观察。hdfs-site.xml 文件配置的与之前一样，不过在集群环境下，还需要配置集群所对应的 NameService。NameService 是在虚拟集群服务名称。由于配置比较多，所以将每一个配置的具体含义，都直接添加到了配置的上面。

```xml
<configuration>
    <!--指定 hdfs 的 nameservice 为 cluster，需要和 core-site.xml 中的保持一致 -->
    <property>
        <name>dfs.nameservices</name>
        <value>cluster</value>
    </property>
    <!-- cluster 下面有多个 NameNode，分别取名为 nn1，nn2，nn3，也可以取其他名称注意，namenodes 后缀为 cluster 即之前配置的名称-->
    <property>
        <name>dfs.ha.namenodes.cluster</name>
        <value>nn1,nn2,nn3</value>
    </property>
    <!--配置每一个 NameNode 的 rpc 通信地址-->
    <property>
        <name>dfs.namenode.rpc-address.cluster.nn1</name>
        <value>server101:8020</value>
    </property>
    <property>
        <name>dfs.namenode.rpc-address.cluster.nn2</name>
        <value>server102:8020</value>
    </property>
    <property>
        <name>dfs.namenode.rpc-address.cluster.nn3</name>
        <value>server103:8020</value>
    </property>
    <!--配置每一个 NameNode 的 web http 地址-->
    <property>
        <name>dfs.namenode.http-address.cluster.nn1</name>
        <value>server101:9870</value>
    </property>
```

```xml
<property>
    <name>dfs.namenode.http-address.cluster.nn2</name>
    <value>server102:9870</value>
</property>
<property>
    <name>dfs.namenode.http-address.cluster.nn3</name>
    <value>server103:9870</value>
</property>
<!--配置QJM的地址-->
<property>
    <name>dfs.namenode.shared.edits.dir</name>
    <value>qjournal://server101:8485;server102:8485;server103:8485/cluster</value>
</property>
<!--配置QJM日志的目录-->
<property>
    <name>dfs.journalnode.edits.dir</name>
    <value>/app/datas/hadoop/qjm</value>
</property>
<!--配置为自动切换功能打开,需要在core-site.xml文件中配置ZK地址-->
<property>
    <name>dfs.ha.automatic-failover.enabled</name>
    <value>true</value>
</property>
<property>
    <name>dfs.client.failover.proxy.provider.cluster</name>
    <value>org.apache.hadoop.hdfs.server.namenode.ha.ConfiguredFailoverProxyProvider</value>
</property>
<!--配置自动切换的方式-->
<property>
    <name>dfs.ha.fencing.methods</name>
    <value>
        sshfence
        shell(/bin/true)
    </value>
</property>
<!--配置SSH key,注意根据不同的用户名修改目录-->
<property>
    <name>dfs.ha.fencing.ssh.private-key-files</name>
    <value>/home/hadoop/.ssh/id_rsa</value>
</property>
<property>
    <name>dfs.permissions.enabled</name>
    <value>false</value>
</property>
<!-- 配置sshfence隔离机制超时时间 -->
<property>
```

```
            <name>dfs.ha.fencing.ssh.connect-timeout</name>
            <value>30000</value>
        </property>
</configuration>
```

步骤 06 配置 mapred-site.xml。

配置 MapReduce 与之前的配置完全一样，配置调试模式为 YARN 即可。

```
<configuration>
<!-- 指定 mr 框架为 YARN 方式 -->
    <property>
        <name>mapreduce.framework.name</name>
        <value>yarn</value>
    </property>
</configuration>
```

步骤 07 配置 yarn-site.xml。

配置 yarn-site.xml 文件与配置 hdfs-site.xml 相同，需要指定集群的配置环境。为了方便阅读，直接将配置说明添加到配置功能之上即可。

```
<configuration>
<!--配置 RM 高可靠启用-->
    <property>
        <name>yarn.resourcemanager.ha.enabled</name>
        <value>true</value>
    </property>
    <!-- 给 resource manager 取一个任意的名称-->
    <property>
        <name>yarn.resourcemanager.cluster-id</name>
        <value>cluster1</value>
    </property>
    <!--配置 Resourcemanager 个数，hadoop3 以后可以为 3 个以上值部分为任意名称-->
    <property>
        <name>yarn.resourcemanager.ha.rm-ids</name>
        <value>rm1,rm2,rm3</value>
    </property>
    <!--以下配置每一个 RM 的地址-->
    <property>
        <name>yarn.resourcemanager.hostname.rm1</name>
        <value>server101</value>
    </property>
    <property>
        <name>yarn.resourcemanager.hostname.rm2</name>
        <value>server102</value>
    </property>
    <property>
        <name>yarn.resourcemanager.hostname.rm3</name>
        <value>server103</value>
    </property>
    <!--配置每一个 RM 的 http 地址-->
    <property>
        <name>yarn.resourcemanager.webapp.address.rm1</name>
        <value>server101:8088</value>
```

```xml
        </property>
        <property>
            <name>yarn.resourcemanager.webapp.address.rm2</name>
            <value>server102:8088</value>
        </property>
        <property>
            <name>yarn.resourcemanager.webapp.address.rm3</name>
            <value>server103:8088</value>
        </property>
        <!--配置 ZooKeeper 地址-->
        <property>
            <name>yarn.resourcemanager.zk-address</name>
            <value>server101:2181,server102:2181,server103:2181</value>
        </property>
        <property>
            <name>yarn.nodemanager.aux-services</name>
            <value>mapreduce_shuffle</value>
        </property>
        <!--hadoop3 里面必须添加的 classpath-->
        <property>
            <name>yarn.application.classpath</name>
            <value>
                请自行将 hadoop classpath 执行结果添加到这儿
            </value>
        </property>
</configuration>
```

步骤08 配置 workers 文件。

workers 是指定 DataNode 节点的位置。在里面添加主机的名称或是 IP 地址即可，一行一个：

```
server101
server102
server103
```

步骤09 配置 Hadoop 的环境变量。

在 server101 主机，配置 Hadoop 环境变量，并通过 scp 将 profile 文件分发到相同的目录下。

```
$sudo vim /etc/profile
export HADOOP_HOME=/app/hadoop-3.2.2
export PATH=$PATH:$HADOOP_HOME/bin
$ sudo scp /etc/profile root@server102:/etc/
$ sudo scp /etc/profile root@server103:/etc/
```

让环境变量生效，执行以下命令：

```
$source /etc/profile
```

步骤10 复制文件其他主机。

将配置好的 Hadoop 目录和 Hadoop 配置文件，复制到其他主机相同的目录下并使用 scp 命令。由于 share 目录下的 doc 里面都是文档，可以删除这个目录，以加快复制速度。

删除 doc 目录：

```
$ rm -rf /app/hadoop-3.2.2/share/doc
```

复制文件：

```
$ scp -r /app/hadoop-3.2.2  server102:/app/
$ scp -r /app/hadoop-3.2.2  server103:/app/
```

步骤 11 启动 journalnode。

分别在 server101、server102、server103 上执行 JournalNode 启动的工作：

```
$ /app/hadoop-3.2.2/bin/hadoop-daemon.sh start journalnode
```

步骤 12 格式化 HDFS。

在 server101 主机上执行格式化 NameNode 的命令：

```
$ hdfs namenode -format
```

格式化后会根据 core-site.xml 中的 hadoop.tmp.dir 配置生成一个文件，然后将这个文件使用 scp 复制到 server102、server103 机器上的相同目录下。因为都是 NameNode 节点，必须拥有相同的数据文件。格式化成功的标志是在输出的日志中查看是否存在以下语句：

```
Storage directory /opt/hadoop_tmp_dir/dfs/name has been successfully formatted
```

现在将格式化后的 HDFS 目录复制到 server102、server103 主机的相同目录下：

```
$ scp -r /app/datas/hadoop  server102:/app/datas/
$ scp -r /app/datas/hadoop  server103:/app/datas/
```

步骤 13 格式化 ZKFC。

在 server101 上执行格式化 ZKFC 的代码如下：

```
$ hdfs zkfc -formatZK
Successfully created /hadoop-ha/cluster in ZK.
```

在格式化完成以后，通过 zkCli.sh 登录 ZooKeeper 并查看目录列表，将显示一个 hadoop-ha 的目录，表示格式化成功：

```
[zk: localhost:2181(CONNECTED) 0] ls /
[zookeeper, hadoop-ha]
```

步骤 14 启动 HDFS。

在 server101 上启动 HDFS 即 NameNode，此命令会根据配置信息同时将 server102、server103 上的 NameNode 一并启动：

```
$ /app/hadoop-3.2.2/sbin/start-dfs.sh
```

现在就可以通过 haadmin 命令查看整个集群的情况，此命令会输出集群中 NameNode 节点的活动状态，其中 active 为活动节点，standby 为备份节点。

```
$ hdfs haadmin -getAllServiceState
server101:8020        active
```

```
server102:8020         standby
server103:8020         standby
```

步骤15 启动 YARN。

在 server101 上执行启动 YARN 的命令，此命令会根据配置信息同时启动 server102 和 server103 上的 ResouceManager 进程。

```
$ /app/hadoop-3.2.2/sbin/start-yarn.sh
```

在启动完成以后，根据之前的配置列表，分别检查每一个主机上的服务是否都已经启动。如果没有请查看日志错误。

启动完成以后，使用 rmadmin 命令查看 ResourceManager 的集群状态。与 NameNode 一样，结果中 active 表示活动节点，standby 为准备节点。

```
$ yarn rmadmin -getAllServiceState
server101:8033         active
server102:8033         standby
server103:8033         standby
```

步骤16 验证高可靠。

通过浏览器访问地址 http://server102:9870，可以查看每一台 HDFS 的信息，如图 7-3 所示。由于 server102、server103 是准备节点，所以显示的状态为 standby。

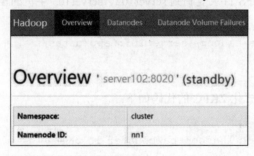

图 7-3

浏览器访问地址 http://server103:9870，页面如图 7-4 所示。

通过图 7-3 可以看出，当前 server101 的 NameNode 为 active。

使用浏览器访问 http://server101:9870，页面如图 7-5 所示。

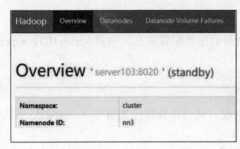

图 7-4   图 7-5

也可以通过以下命令，检查 NameNode 和 ResourceManager 的状态：

```
$ hdfs haadmin -getServiceState nn1
active
$ hdfs haadmin -getServiceState nn2
standby
$ yarn rmadmin -getServiceState rm1
active
$ yarn rmadmin -getServiceState rm2
standby
```

现在我们 kill 掉 active 的 NameNode，即 kill 掉 nn1：

```
$ kill -9 <pid of NN>
```

然后再检查状态，此时 server102 上的 NameNode 变成了 active：

```
$ hdfs haadmin -getServiceState nn2
Active
```

也可以通过 haadmin 命令查看节点状态：

```
$ hdfs haadmin -getAllServiceState
server101    failed to connected
server102    active
server103    standby
```

手动启动那个挂掉的 NameNode，即 nn1，然后再检查状态，它已经成为 standby 了：

```
$ ./hadoop-daemon.sh start namenode
$ hdfs haadmin -getServiceState nn1
standby
```

使用同样的方式可以验证 ResourceManager 是否可以自动实现容灾切换。

注意：（1）在集群完成以后，建议执行一个 MapReduce 测试，如 wordcount；（2）Hadoop 的高可靠集群每一次启动相对比较麻烦。但配置成功以后，下次启动就相对比较简单了。对于上面的示例而言，再次启动只要在 server101 主机上执行 ./start-dfs.sh 和 ./start-yarn.sh 即可。

7.6 用 Java 代码操作集群

用 Java 客户端操作集群开发 HDFS，必须指定 NameService 的配置信息。以下代码示例用于显示 HDFS 上的文件和目录。

【代码 7-1】HdfsAccess.java

```
1.   package org.hadoop.ha;
2.   public class HdfsAccess extends Configured implements Tool {
3.       @Override
4.       public int run(String[] args) throws Exception {
5.           System.setProperty("HADOOP_USER_NAME", "hadoop");
```

```
6.      Configuration conf = getConf();
7.      conf.set("fs.defaultFS", "hdfs://cluster");
8.      conf.set("dfs.nameservices", "cluster");
9.      conf.set("dfs.ha.namenodes.cluster", "nn1,nn2,nn3");
10.     conf.set("dfs.namenode.rpc-address.cluster.nn1", "server101:8020");
11.     conf.set("dfs.namenode.rpc-address.cluster.nn2", "server102:8020");
12.     conf.set("dfs.namenode.rpc-address.cluster.nn3", "server103:8020");
13.     conf.set("dfs.client.failover.proxy.provider.cluster",
14.             "org.apache.hadoop.hdfs.server.namenode.ha.ConfiguredFailoverProxyProvider");
15.     FileSystem fs = FileSystem.get(conf);
16.     FileStatus[] statuses = fs.listStatus(new Path("/"));
17.     for (FileStatus f : statuses) {
18.         System.err.println(f.getPath().toString());
19.     }
20.     fs.close();
21.     return 0;
22. }
23. public static void main(String[] args) throws Exception {
24.     int code = ToolRunner.run(new HdfsAccess(), args);
25.     System.exit(code);
26. }
27. }
```

以下代码是保存一个文件到高可靠集群的示例。

【代码 7-2】HASaveHdfs.java

```
1.  package org.hadoop.ha;
2.  public class HASaveHdfs extends Configured implements Tool {
3.      @Override
4.      public int run(String[] args) throws Exception {
5.          System.setProperty("HADOOP_USER_NAME", "hadoop");
6.          Configuration conf = getConf();
7.          conf.set("fs.defaultFS", "hdfs://cluster");
8.          conf.set("dfs.nameservices", "cluster");
9.          conf.set("dfs.ha.namenodes.cluster", "nn1,nn2,nn3");
10.         conf.set("dfs.namenode.rpc-address.cluster.nn1", "server101:8020");
11.         conf.set("dfs.namenode.rpc-address.cluster.nn2", "server102:8020");
12.         conf.set("dfs.namenode.rpc-address.cluster.nn3", "server103:8020");
13.         conf.set("dfs.client.failover.proxy.provider.cluster",
14.                 "org.apache.hadoop.hdfs.server.namenode.ha.ConfiguredFailoverProxyProvider");
15.         FileSystem fs = FileSystem.get(conf);
16.         OutputStream out = fs.create(new Path("/test/c.txt"));
```

```
17.         out.write("Hello 中文".getBytes());
18.         out.close();
19.         fs.close();
20.         return 0;
21.     }
22.     public static void main(String[] args) throws Exception {
23.         int code = ToolRunner.run(new HASaveHdfs(), args);
24.         System.exit(code);
25.     }
26. }
```

至此，你已经掌握如何部署 Hadoop 的高可靠集群了，并学会如何通过 Java 代码访问高可靠集群。

7.7　本章小结

本章主要讲解了以下内容：

- ZooKeeper 集群配置。
- Hadoop 借助 ZooKeeper 实现 HA 高可靠。
- ZKFC 用于实时向 ZooKeeper 汇报 NameNode 的状态，一旦 NameNode 不可用，就会自动进行切换。
- JournalNode 用于实时同步两个 NameNode 之间的日志数据。
- 可以使用 hdfs haadmin 查看 NameNode 的集群状态。可以使用 yarn rmadmin 查看 ResourceManager 的集群状态。

第 8 章

数据仓库 Hive

本章主要内容：

- Hive 的体系结构和特点。
- Hive 的命令。
- Hive 视图。
- Hive 表分区。
- Hive 之 UDF 编程。
- Hive 之 metastore server。

Hive 是基于 Hadoop 的一个数据仓库，可以将结构化的数据文件映射为一张数据库表，并提供 SQL 查询功能。Hive 在执行 SQL 时会将 SQL 语句转换为 MapReduce 任务运行。其优点是学习成本低，可以通过类似于 SQL 的语句快速实现 MapReduce 的开发，不必编写专门的 MapReduce 应用程序，十分适合离线数据的统计分析。

8.1 Hive 简介

Hive 是建立在 Hadoop 上的数据仓库基础构架。它提供了一系列的工具，可以用来进行数据抽取、转换、加载（Extract-Transform-Load，ETL），这是一种可以对存储在 Hadoop 中的大规模数据进行查询和分析的机制。Hive 定义的类似于 SQL 的查询语言，称为 HQL，它允许熟悉 SQL 的用户查询数据。同时，这个语言也允许熟悉 MapReduce 开发者开发自定义的 Mapper 和 Reducer，以处理内建的 Mapper 和 Reducer 无法完成的复杂的分析工作。

Hive3 运行在 Hadoop 3 之上。以下是官方的发布说明，其中约定了 Hive 什么版本可以运行在 Hadoop 的什么版本之上。

17 January 2021: release 2.3.8 available

```
This release works with Hadoop 2.x.y You can look at the complete JIRA change
log for this release.
26 August 2019: release 3.1.2 available
This release works with Hadoop 3.x.y. You can look at the complete JIRA change
log for this release.
```

Hive 的特点如下：

- 对仓库中的数据进行分析和计算。
- 建立在 Hadoop 之上。
- 一次写入、多次读取。
- Hive 是 SQL 语句分析引擎，将 SQL 语句转换成 MapReduce 并在 Hadoop 上的执行。
- Hive 表对应的是 HDFS 的文件夹。
- Hive 表的数据对应的是 HDFS 的文件。

Hive 体系结构如图 8-1 所示。

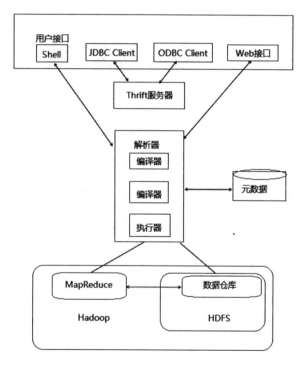

图 8-1

Hive 的主要组成部分分为以下几个部分。

1. 用户接口

用户接口主要有三个：CLI、Client 和 WUI。其中最常用的是 CLI，CLI 启动的时候，会同时启动一个 Hive 副本。Client 是 Hive 的客户端，用户可以通过它连接到 Hive Server。在启动 Client 模式的时候，需要指定 Hive Server 所在的节点，并且在该节点上已经启动 Hive Server。WUI 是通过浏览器访问 Hive。

2. 元数据存储

Hive 将元数据存储在数据库中，如 MySQL、Derby。Hive 中的元数据包括表的名字、表的列和分区及其属性、表的属性（是否为外部表等）、表的数据所在目录等。

目前已经支持 Hive 元数据存储的数据库包括：

- MySQL：5.6.17 及更新的版本。
- PostgreSQL：9.1.13 及更新的版本。
- Oracle：11g 及更新的版本。
- MS SQL Server：2008 R2 及更新的版本。

3. 解释器、编译器、优化器、执行器

解释器、编译器、优化器完成 HQL 查询语句从词法分析、语法分析、编译、优化以及查询计划的生成。生成的查询计划存储在 HDFS 中，并在随后由 MapReduce 调用执行。

Hive 的数据存储在 HDFS 中，执行 Hive 的大部分 SQL 将会引发 MapReduce 的执行。

8.2 Hive3 的安装配置

到本书完稿时，Hive 已经更新到 3.1.2 版本。使用 Hive 的前置条件：启动 HDFS，启动 YARN。如果仅有一台服务器，并已经安装了 Hadoop 伪分布式及 ZooKeeper 即可以开始 Hive 的学习。

步骤 01 下载 Hive3。

下载 Hive 网址为 https://mirrors.bfsu.edu.cn/apache/hive/hive-3.1.2/apache-hive-3.1.2-bin.tar.gz。

步骤 02 解压缩。

可以将 Hive 解压缩到任意的目录下，此处为了方便操作，将 Hive 解压缩到安装集群 Hadoop 相同目录下。

```
$ tar -zxvf ~/apache-hive-3.1.2-bin.tar.gz -C /app/
```

步骤 03 初始化数据库。

Hive 的运行同样需要元数据的支持，Hive 元数据可以保存到关系型数据库中。默认情况下，保存到 Derby 数据库中。如果需要快速执行，可以直接选择使用 Derby 数据库。使用 Derby 数据库的缺点是，不能同时使用两个 Hive CLI 共同操作 Hive 数据仓库，因为 Derby 这种类型的数据库，不支持多用户同时登录。所以，我们需要将元数据保存到类似于 MySQL 的关系型数据库中，具体将在后面的章节讲解。

8.2.1 使用 Derby 数据库保存元数据

默认情况下，Hive 使用 Derby 数据库作为元数据管理数据库。使用 Derby 数据库，会在当前目录下创建 metastore_db 目录，即为 Derby 数据库的目录。与 Hive2 和 Hive1 不同，Hive3 在使用 Derby

数据库之前，必须先执行初始化命令，一个 Derby 数据库只允许一个 Hive 客户端登录。

以下初始化 Derby 数据库（请确定已经启动 HDFS）：

```
$ /app/hive-3.1.2/bin/schematool -dbType derby -initSchema
```

成功后显示：

```
Initialization script completed
schemaTool completed
```

初始化成功以后，就可以使用 hive 命令登录 Hive 客户端：

```
$/app/hive-3.1.2/bin/hive
Hive>
```

如果显示了 Hive>则表示登录 Hive 的命令行成功，此时就可以使用 Hive SQL 操作数据了。Hive SQL 使用了类似于 MySQL 的 SQL 语法，因此，有过 SQL 编程经验的朋友应该可以快速上手。

显示当前有多少数据库：

```
Hive>show databases;
Default
```

创建一个数据库，会在 HDFS 的 user/hive/warehouse/下创建一个 xx.db 的数据库目录。注意，默认情况下，Hive 的数据库保存在 hdfs://xxx:8020/user/hive/warehouse 目录下。

```
Hive>create database one;
```

查看 Hive 数据库：

```
Hive>show databases;
Default
One
```

查看 HDFS 目录：

```
$ hdfs dfs -ls /user/hive/warehouse
one.db
```

8.2.2 使用 MySQL 数据库保存元数据

使用 MySQL 保存元数据的优点就是多个 Hive CLI 客户端可以同时登录，并操作相同的数据。连接到的 MySQL 数据库来保存元数据，需要将 mysql-connector.jar 的驱动文件放到 Hive3 的 lib 目录下，然后配置 hive-site.xml 文件。

配置 HIVE_HOME/config/hive-site.xml 文件，将 hive-site.template.xml 文件重命名为 hive-site.xml 文件，然后输入以下内容：

```xml
<?xml version="1.0" encoding="UTF-8" standalone="no"?>
<?xml-stylesheet type="text/xsl" href="configuration.xsl"?>
<configuration>
    <property>
        <name>javax.jdo.option.ConnectionDriverName</name>
        <value>com.mysql.cj.jdbc.Driver</value>
```

```xml
        </property>
        <property>
            <name>javax.jdo.option.ConnectionURL</name>
            <value>jdbc:mysql://127.0.0.1:3306/hive?characterEncoding=UTF-8&amp;serverTimezone=Asia/Shanghai&amp;useSSL=false
            </value>
        </property>
        <property>
            <name>javax.jdo.option.ConnectionUserName</name>
            <value>root</value>
        </property>
        <property>
            <name>javax.jdo.option.ConnectionPassword</name>
            <value>123456</value>
        </property>
</configuration>
```

与使用 Derby 数据库一样，首先需要初始化数据库操作，所以先执行初始化数据库命令：

```
$ /app/hive-3.1.2/bin/schematool -dbType mysql -initSchema
Initialization script completed
schemaTool completed
```

在初始化完成以后，会在 MySQL 的 Hive 数据库中查看到已经初始化的数据表：

```
mysql> show tables;
+-------------------------------+
| Tables_in_hive                |
+-------------------------------+
| AUX_TABLE                     |
| BUCKETING_COLS                |
| CDS                           |
| COLUMNS_V2                    |
| COMPACTION_QUEUE              |
| COMPLETED_COMPACTIONS         |
| COMPLETED_TXN_COMPONENTS      |
| CTLGS                         |
| DATABASE_PARAMS               |
| DBS                           |
| DB_PRIVS                      |
...
```

创建完成以后，就可以使用 hive 命令登录 Hive 的 CLI 操作数据了。

```
[hadoop@server201 app]$ hive
hive> show databases;
OK
Default
```

查看 MySQL 数据库中的表如下所示：

```
hive> show databases;
OK
```

```
default
one
```

步骤04 配置环境变量（可选）。

建议将 Hive 的 bin 目录配置到环境变量，以更加方便地使用 hive 命令。

```
export HIVE_HOME=/app/hive-3.1.2
export PATH=$PATH:$HIVE_HOME/bin
```

现在，就可以不用进入 Hive 的安装目录去登录 Hive 客户端了。可以在任意目录下，执行 hive 登录 Hive 的命令行模式：

```
$ hive
```

在登录 Hive 命令行以后，执行类似于 MySQL 的命令 show databases，即可显示当前所有的数据库：

```
hive> show databases;
OK
default
Time taken: 1.391 seconds, Fetched: 1 row(s)
```

上面的输出结果中存在一个默认的数据库。还可以执行 show tables 显示默认数据库下的所有表：

```
hive> show tables;
```

在启动 Hive 以后，会在 Hadoop HDFS 上出现：/user/hive/warehouse 目录，这就是用于保存 Hive 数据的目录。到此为止，你的 Hive 已经可以运行了。

步骤05 查看 hive 命令的帮助。

同样地，通过--help 参数可以查看所有 Hive 的帮助信息。

```
$ hive --help
Usage ./hive <parameters> --service serviceName <service parameters>
```

其中，parameters 可以是：

```
--auxpath : Auxiliary jars
--config : Hive configuration directory
--service : Starts specific service/component. cli is default
```

而 serviceName 默认为 cli。

如需显示版本可以在--service 后面添加 version：

```
$ hive --service version
$ Hive-3.1.2
```

更多的帮助请自行查看命令行说明。

8.3 Hive 命令

Hive 的很多命令类似于 SQL 命令。但有些命令与 SQL 存在一些差异，Hive 拥有自己独特的 SQL 语法。

1. 创建一个数据库

Hive 中已经有了一个默认的数据库——default，现在你可以创建一个自己的数据库：

```
hive> create database one;
```

在创建完成这个数据库以后，就可以使用 show databases 显示 Hive 下的所有数据库：

```
hive> show databases;
OK
default
one
```

2. 创建一个表

现在可以使用 use one 命令进入 one 数据库，并在这个数据库下创建一个表：

```
hive> create table stud(id int,name string);
```

在创建表以后，查看 HDFS 目录，会发现目录/user/hive/warehouse/one.db/stud。这就是刚才创建的表。

默认情况下，创建的表只会在元数据库（即 MySQL 或 Derby）保存一个数据结构，向里面写入数据时，默认会使用"^A"作为列的分割符。建议使用 row delimited 来指定分割符号。

3. 显示表结构

```
Hive> desc stud;
Hive>desc formatted stud;
```

还可以使用 show create table 表名来显示表的结构：

```
1.  hive> show create table stud;
2.  OK
3.  CREATE TABLE `stud`(
4.      `id` int,
5.      `name` varchar(30)
6.  ) ROW FORMAT SERDE
7.    'org.apache.hadoop.hive.serde2.lazy.LazySimpleSerDe'
8.    STORED AS INPUTFORMAT
9.    'org.apache.hadoop.mapred.TextInputFormat'
10.   OUTPUTFORMAT
11.   'org.apache.hadoop.hive.ql.io.HiveIgnoreKeyTextOutputFormat'
12.   LOCATION
13.   'hdfs://server201:8020/user/hive/warehouse/one.db/stud'
14.   TBLPROPERTIES (
15.   'transient_lastDdlTime'='1501079369')
```

```
16.   Time taken: 0.886 seconds, Fetched: 13 row(s)
```

上面显示的表结构中，第 9 行代码用于声明这个文件的读取类。第 10~11 行代码声明输出数据类。第 12~13 行代码是操作目录。第 14~15 行代码为这个表最后修改的时间，当前显示为创建的时间。

甚至可以向这个表中保存数据（不建议这么做），由于 Hive 中的表对应的是文件，所以向 Hive 表中保存数据就是向 HDFS 文件中保存数据。

向表中写入数据，虽然 Hive SQL 支持 insert 语句，但在真实的环境下，数据一般是从外部文件直接导入或是关联到 Hive 仓库的。这里执行 insert 语句只是为了一个测试：

```
hive> insert into stud values(1,'SomeValue');
```

当执行 insert 向表中写入数据时，默认会执行 Mapper 任务向 HDFS 文件中写入数据。所以，你会发现上面的语句执行了一个 Mapper 任务。再次声明，一般情况下，我们不会执行 insert 写入数据，而是从外部（非 HDFS 文件）或是内部（HDFS 文件）中导入数据。

现在可以执行 select 查询表中的数据：

```
hive> select * from stud;
OK
1    SomeValue
```

从代码中可以发现，已经存在一行数据了。再查询 HDFS 文件系统上的数据：

```
hive> dfs -cat /user/hive/warehouse/opt.db/stud/*;
1SomeValue
```

在 Hive 客户端命令行上，可以直接执行 dfs 命令，类似于执行 HDFS dfs 命令，只是省去了 HDFS。通过上面的结果可以看出，里面已经写入了 1SomeValue 一行数据，且并没有分隔符号。现在我们将里面的文件 000000_0 下载到本地：

```
hive> dfs -get /user/hive/warehouse/opt.db/stud/* /home/hadoop/;
```

现在可以通过 vim 查看下载的文件：

```
$vim 000000_0
1^ASomeValue
```

可见，1 与 Jack 之间是使用 "^A" 作为分隔符号的。现在我们可以在创建一个表时，指定数据之间的分隔符号。

4. 创建一个表，并指定分隔符号

```
hive> create table person(
    >    id int,
    >    name varchar(30)
    > )
    > row format delimited fields terminated by '\t';
```

注意上面的语句中最后的 ";"（分号），如果没有 ";" 就像 SQL 语句没有结束一样，所以在 Hive 中 ";" 也是语句结束的标记。上例中，设置\t（制表）符号为字段数据之间分隔的标记。

现在可以再写入一行数据，测试一下数据在 HDFS 文件中的分隔符号：

```
hive> insert into person(id,name) values(100,'Mary');
```

5. 查看数据的分隔符号

```
hive> dfs -cat /user/hive/warehouse/opt.db/person/*;
100    Mary
```

通过上面的结果可以看出，字段之间已经使用\t进行了分隔。

6. load data 上传本地文件

load data 命令用于上传一个本地文件到 Hive 的一个表中。其中，local 参数用于加载本地磁盘上的一个文件。如果没有 local 参数，则为加载 HDFS 文件上的文件到 Hive 的表中。

默认情况下，程序处理的流程是先使用 MapReduce 将原始数据处理成具有一定格式的文本文件，然后再通过 load data 命令将数据加载到 Hive 的表中，然后使用 HQL 进行数据分析。

如要将一个文件中的数据导入到上述的 person 表中，由于 person 表中字段之间的数据是用\t分隔的。所以，可以先通过 vim 创建一个文件，并用\t分隔里面的数据：

```
$ vim person.txt
101    Jack
102    Mary
103    Mark
104    Alex
```

现在使用 load data 命令，将数据导入到 person 表中去：

```
hive> load data local inpath '/home/hadoop/person.txt' into table person;
```

在执行上面的导入语句以后，会在 HDFS 上发现 person.txt 这个文件，这个文件所在的目录为 /user/hive/warehouse/opt.db/person/person.txt。

现在查询里面的数据：

```
hive> select * from person;
OK
100    Mary
101    Jack
102    Mary
103    Mark
104    Alex
```

也可以上传一个 HDFS 文件到 person 表中，使用 load data 命令，不添加 local 即是从 HDFS 上加载数据。如先将某个文件上传到 HDFS 上：

```
hive> dfs -put /home/hadoop/person.txt /person.txt;
```

再使用 load data 将 HDFS 上的文件载入到 Hive 的表中。需要注意，上传完成以后，HDFS 上的文件会被移动到 Hive 中，即会删除原目录下的文件：

```
hive> load data inpath '/person.txt' into table person;
```

7. 执行 MapReduce 任务

对数据进行统计 count，会执行一个 MapReduce 任务。如以下代码：

```
hive> select count(*) from person;
OK
15
```

执行上面的查询语句后会发现一个完整的 MapReduce 过程，并最终直接将结果输出到控制台。也可以将计算的结果输出到本地文件中。注意，下面例子中输出的 count 是一个目录，里面的文件才是输出的结果数据：

```
hive> insert overwrite local directory "/home/hadoop/count"
> select count(*) from person;
```

也可指定导出的数据的分隔符号：

```
Hive>insert overwrite local directory '/home/hadoop/out001'
>row format delimited
>fields terminated by '\t'
select id,name from t1;
```

更可以将计算的结果输出到 HDFS 上去，去掉 local 参数即可。同样地，/count 是一个目录，里面文件中的数据才是我们需要的结果。

```
hive> insert overwrite  directory '/count' select count(1) from person;
```

执行一个过滤排序的查询，会执行 MapReduce 任务：

```
hive> select * from person where id>102 order by id;
Total MapReduce CPU Time Spent: 3 seconds 440 msec
OK
104     Alex
104     Alex
104     Alex
105     Mark
105     Mark
105     Mark
```

到此为止，你已经执行了 Hive 的一些命令。除 "select *" 命令不会生成 MapReduce 之外，其他命令都会生成 MapReduce 任务。可见，Hive 极大地简化了 MapReduce 的开发。

8.4　Hive 内部表

默认情况下，使用 create table 创建的表都是 Hive 的内部表。创建的内部表保存在 /user/hive/warehouse 目录下。执行 drop table 内部表时，会将数据及数据文件全部删除。

在元数据管理中，内部表在 TBLS 元数据中显示为：MANAGED_TABLE，上一节所创建的表都是内部表。

8.5 Hive 外部表

如果数据已经存在于 HDFS 上，则可以通过创建外部表的方式与 HDFS 上的数据建立关系。默认通过 create table 创建表为内部表（managed_table）。在元数据 tbls 表中有一列 TBL_TYPE，如果值为 MANAGED_TABLE 则为内部表。

通过 create external table 可以创建一个外部表。在创建时，需要指定 HDFS 上的一个目录，创建外部表与目录的关系：

```
hive> create external table ext_person(id bigint,name string)
    > row format
    > delimited fields terminated by '\t'
    > location '/test/person';
OK
Time taken: 0.149 seconds
```

在上面的代码中，name 可以使用 varchar(N)，也可以使用 string。注意，最后的 location '/test/person' 用于指定 HDFS 的目录，此目录下包含多个 txt 类型的文件，里面保存了以\t 分隔的 id 和 name 数据。现在查看 tbls 表的 tbl_type 列，值为 EXTERNAL_TABLE 即为外部表。

查询外部表就像是查询内部表一样，可以获取数据结果：

```
hive> select * from ext_person;
OK
101 Jack
102 Mary
103 Mark
104 Alex
```

外部表创建后，TBLS 元数据表中保存了外部表的信息：table_type=EXTERNAL_TABLE。

使用 drop table 删除表之后，外部表所引用的 HDFS 目录不会被删除，当然 HDFS 目录下的相关数据也不会被删除。

8.6 Hive 表分区

Hive 表分区的原因如下：

（1）在 Hive select 查询中一般会扫描整个表内容，会消耗很多时间做没必要的工作。有时候只需要扫描表中关心的一部分数据，因此建表时引入了 partition（分区）的概念。

（2）分区表指的是在创建表时指定了 partition 分区空间的表。

（3）如果需要创建有分区的表，需要在 create 表的时候传入可选参数 partitioned by 关键字。

8.6.1 分区的技术细节

（1）一个表可以拥有一个或者多个分区，每个分区以文件夹的形式出现在 HDFS 文件系统的目录中。

（2）表和列名不区分大小写。

（3）分区是以字段的形式在表结构中存在，通过 describe table 命令可以查看到字段的存在，但是该字段不存放实际的数据内容，仅仅是分区的表示。

（4）建表的语法（建分区可参见 PARTITIONED BY 参数）为：

```
CREATE [EXTERNAL] TABLE [IF NOT EXISTS] table_name [(col_name data_type [COMMENT col_comment], ...)] [COMMENT table_comment] [PARTITIONED BY (col_name data_type [COMMENT col_comment], ...)] [CLUSTERED BY (col_name, col_name, ...) [SORTED BY (col_name [ASC|DESC], ...)] INTO num_buckets BUCKETS] [ROW FORMAT row_format] [STORED AS file_format] [LOCATION hdfs_path]
```

（5）分区建表分为 2 种，一种是单分区，也就是说在表文件夹目录下只有一级文件夹目录；另外一种是多分区，表文件夹下出现多文件夹嵌套模式。

单分区建表语句中使用 partitioned by 即可，如以下语句通过 partitioned by 创建了一个分区 grade（年级），将会在/user/hive/warehouse/stud 目录下创建多个不同的分区目录，如 01 年级的数据就会保存到/user/hive/warehouse/stud/01 下。

```
hive> create table stud(id int,name string) partitioned by (grade string)
    > row format delimited fields terminated by '\t';
```

现在可以使用 insert 插入一些数据，虽然我们不建议这么做，但这样做可以让你快速了解分区的存储形式：

```
hive> insert into stud(id,name,grade) values(2,'Mike','2002');
hive> insert into stud(id,name,grade) values(1,'Jack','2001');
```

以上两个语句会开启 MR 程序，写入两行数据。插入数据以后，查看 HDFS 的目录结构即可以看到 2001 和 2002 两个子目录。为了显示表头，可以设置 Hive 显示表头：

```
hive> set hive.cli.print.header=true;
```

查询数据：

```
hive> select * from stud;
OK
stud.id  stud.name   stud.grade
1        Jacsk       2001
2        Mike        2002
```

查看 HDFS 上的数据可见分区，最后形成了目录：

```
/user/hive/warehouse/stud/grade=2001
/user/hive/warehouse/stud/grade-2002
```

导入数据时可以指定分区值，只要在语句最后添加 partition(..)，即可将导入的数据添加到指定的分区：

```
hive> load data local inpath '/root/stud.txt' into table stud partition (grade='2008');
```

创建多个分区：

以下示例在表 students 中创建 id、name 和 major（专业）以及 grade（年级）四个列。其中 major、grade 为分区数据，即根据专业、年级对学生信息进行分区：

```
hive> create table students(id int,name string)
> partitioned by (major string,grade string)
> row format delimited fields terminated by '\t';
```

上表创建两个分区，major 按专业分区，grade 再按年级进行分区。

上表是双分区表，即在表结构中新增加了 major 和 grade 两列。此时，保存到 HDFS 上的文件系统显示结果如下所示。注意，只有在保存了数据以后，才会存在以下结构，且会根据保存数据的不同，分区数据会显示为不同的目录。

```
/user/hive/warehouse/opt.db/students/major=Java/grade=2017
```

（6）添加分区是指在创建表时没有创建分区列，通过修改表的方式，添加新的分区信息。

```
ALTER TABLE table_name ADD partition_spec [ LOCATION 'location1' ] partition_spec [ LOCATION 'location2' ] ... partition_spec: : PARTITION (partition_col = partition_col_value, partition_col = partiton_col_value, ...)
```

用户可以用 ALTER TABLE ADD PARTITION 来向一个表中增加分区。当分区名是字符串时加引号。比如：

```
hive> alter table student add partition(major='Java');
```

（7）删除分区数据。

```
ALTER TABLE table_name DROP partition_spec, partition_spec,...
```

用户可以用 ALTER TABLE DROP PARTITION 来删除分区及里面的数据。分区的元数据和数据将一并被删除。比如：

```
hive> alter table student drop partition(major="Java");
```

（8）数据加载进分区表中语法。

将已有数据，保存到指定分区中。它的语法是：

```
LOAD DATA [LOCAL] INPATH 'filepath' [OVERWRITE] INTO TABLE tablename [PARTITION (partcol1=val1, partcol2=val2 ...)]
```

现在加载数据到指定的分区中：

```
hive> load data local inpath '/home/hadoop/person.txt' into table students
> partition(major='Java',grade='2017');
```

当数据被加载至表中时，不会对数据进行任何转换。Load 操作只是将数据复制至 Hive 表对应的位置。数据加载时在表下自动创建一个目录，文件存放在该分区下。

当查询保存到分区中的数据时，数据的内部与加载之前的数据完全一样。比如：

```
hive> dfs -cat \
 /user/hive/warehouse/opt.db/students/major=Java/grade=2017/*;
101     Jack
...
```

但当执行查询时,会将分区的数据当作表的一部分查询出来。比如:

```
hive> select * from students;
OK
101 Jack      Java      2017
102 Mary      Java      2017
103 Mark      Java      2017
104 Alex      Java      2017
```

(9) 基于分区的查询语句。

现在我们可以使用 load data 导入更多的数据,其后在查询语句中使用分区的字段作为查询条件。

```
hive> select * from students where major='Oracle';
OK
301 Jack      Oracle    2017
302 Mary      Oracle    2017
303 Alex      Oracle    2017
304 Smith     Oracle    2017
```

(10) 查看分区语句。

可以通过 show partitions 查看一个表的分区信息:

```
hive> show partitions students;
OK
major=Java/grade=2017
major=Oracle/grade=2017
Time taken: 0.235 seconds, Fetched: 2 row(s)
```

8.6.2 分区示例

(1) 在 Hive 中,表中的一个 Partition 对应于表下的一个目录,所有的 Partition 的数据都存储在对应的目录中。

(2) 总的说来,partition 就是辅助查询、缩小查询范围、加快数据的检索速度和对数据按照一定的规格和条件进行管理。

以下示例创建一个学生表,表中拥有三个字段 id、name 和 classname(班级名称),并以 classname 作为分区:

```
hive> create table students(id bigint,name string)
    > partitioned by (classname string)
    > row format delimited fields terminated by '\t';
```

表创建成功以后,数据保存到 HDFS 的/user/hive/warehouse/students 目录下。

现在导入一些数据到 students 表中去,先使用 vi 编辑一个文本文件,里面的内容如下:

```
vi studs.txt
101     Jack
102     Mary
...
```

然后使用 load data 将数据保存到 students 表中去,由于 students 表是有分区的,所以在 load data

时必须指定分区信息，注意语句最后的 partition(classname="Java")：

```
hive> load data local inpath '/home/mrchi/studs.txt'
> into table students partition(classname="Java");
```

可以多在表中 load 一些数据，并指定不同的分区信息：

```
hive> load data local inpath '/home/hadoop/studs.txt'
>into table students partition(classname="oracle");
```

在 load 数据时，可以通过 overwite 关键字重新设置某个分区下的所有数据：

```
hive> load data local inpath '/home/wangjian/studs2.txt'
>overwrite into table students partition(classname="Oracle");
```

现在查看 HDFS 上的目录结构，具体如下：

```
hive> dfs -ls /user/hive/warehouse/students;
Found 2 items
/user/hive/warehouse/opt.db/students/classname=Java
/user/hive/warehouse/opt.db/students/classname=Oracle
```

由此可见，在 students 目录中出现了两个子目录，分别为 classname=Java 和 classname=Oracle，即为两个分区目录。

现在查询所有数据：

```
hive> select * from students;
101  Jack     Java
102  Mary     Java
301  Jack     Oracle
302  Mary     Oracle
303  Alex     Oracle
```

显示 partition 的信息：

```
hive> show partitions students;
classname=Java
classname=Oracle
```

创建具有多个分区的表，比如某个班级是哪一年级的哪一个专业，以下即指定两个分区信息：

```
hive> create table classes(id bigint,name string)
  > partitioned by(grade string,major string)
  > row format delimited fields terminated by '\t';
```

导入数据，首先使用 vim 编辑一个文件如 cls1.txt：

```
1     Java1
2     Java2
```

然后再做导入：

```
hive> load data local inpath '/home/hadoop/cls1.txt'
>overwite into table classes partition(grade='2016',major='computer');
```

导入数据以后，会出现两层目录结构：

```
/user/hive/warehouse/classes/grade=2016/major=computer
```

导入更多数据，并查询：

```
hive> load data local inpath '/home/hadoop/cls2.txt'
>overwrite into table classes partition(grade='2017',major='ui');
hive> select * from classes;
1    Java1    2016    computer
2    Java2    2016    computer
9    UI1      2017    ui
10   UI2      2017    ui
```

添加一个新的分区，如添加一个专业信息，即哪一个班级，属于哪一个专业：

```
hive> alter table classes add partition \
 (grade='2015',major='computer');
```

添加一个新的分区，是指在当前表的目录下创建分区目录出来的过程。本质上，不是修改表结构，而是添加数据目录的过程。

添加一个专业信息执行完成以后，新的 HDFS 目录结构为：

```
/user/hive/warehouse/classes/grade=2015/major=computer
```

删除一个分区：

```
hive> alter table classes drop \
partition(grade='2015',major='computer');
```

8.7 查询示例汇总

创建一个表，并指定列的分割方式：

```
hive> create table person(id string,name string,age int,sex string)
    > row format delimited
    > fields terminated by '\t';
```

创建一个表，并指定存储格式为 textfile：

```
hive> create table car(id string,name string,price double,pid string)
    > row format delimited
    > fields terminated by '\t'
    > stored as textfile;
```

关联查询：

```
hive> select p.id,p.name,c.id,c.name from person p inner join car c on p.id=c.pid;
```

使用 group by：

```
hive> select count(1),pid from car group by pid;
```

使用 having：

```
hive> select count(1),pid from car group by pid having count(1)>=2;
```

子查询：

```
hive> select * from person where id in(select pid from car group by pid having count(1)>=2);
```

相关子查询：

```
hive> select * from person where (select count(1) from car where person.id=car.pid)=2;
```

8.8　Hive 函数

Hive 内部定义了很多函数，这些函数都是通过 Hive 的 FunctionRegistry 类注册的。在 IDE 中查看此类的源代码，需要添加 Hive Query Language 的依赖：

```xml
<dependency>
    <groupId>org.apache.hive</groupId>
    <artifactId>hive-exec</artifactId>
    <version>3.1.2</version>
</dependency>
```

依赖添加之后，就可以查看 FunctionRegistry 这个类的源代码了，你将会发现大量类似于以下的函数注册代码：

```java
//registry for system functions
private static final Registry system = new Registry(true);
static {
    system.registerGenericUDF("concat", GenericUDFConcat.class);
    system.registerUDF("substr", UDFSubstr.class, false);
    system.registerUDF("substring", UDFSubstr.class, false);
    system.registerGenericUDF("substring_index", GenericUDFSubstringIndex.class);
    system.registerUDF("space", UDFSpace.class, false);
    system.registerUDF("repeat", UDFRepeat.class, false);
    system.registerUDF("ascii", UDFAscii.class, false);
    system.registerGenericUDF("lpad", GenericUDFLpad.class);
    system.registerGenericUDF("rpad", GenericUDFRpad.class);
    system.registerGenericUDF("levenshtein", GenericUDFLevenshtein.class);
    system.registerGenericUDF("soundex", GenericUDFSoundex.class);
...
```

上面的代码，就是向 Hive 系统注册函数。在执行 HQL 语句的过程中，可以使用 Hive 已经定义的函数。可以使用 show functions 查看 Hive 中所有的内部函数：

```
hive> show functions;
...
<
<=
```

```
<=>
...
abs
acos
add_months
aes_decrypt
aes_encrypt
...
```

通过 show functions 可以列出 Hive 内部已经注册的所有函数，可以看到运算符号都已经被注册成了 Hive 的函数。还有 abs、acos 等数学函数和其他更多的函数。

也可以使用"describe function 函数名称;"来查看具体某一个函数的用法。比如我们可以查看一下运算符号"="函数的具体说明：

```
hive> describe function =;
OK
a = b - Returns TRUE if a equals b and false otherwise
Time taken: 0.012 seconds, Fetched: 1 row(s)
```

函数 uuid 的功能如下：

```
hive> describe function uuid;
OK
uuid() - Returns a universally unique identifier (UUID) string.
Time taken: 0.015 seconds, Fetched: 1 row(s)
```

接下来讲解 Hive 中的函数及其示例。

8.8.1 关系运算符号

在 FunctionRegistry 类中，可以找到注册关系运算位的源代码：

```
HIVE_OPERATORS.addAll(Arrays.asList(
    "+", "-", "*", "/", "%", "div", "&", "|", "^", "~",
    "and", "or", "not", "!",
    "=", "==", "<=>", "!=", "<>", "<", "<=", ">", ">="));
```

通过 HIVE_OPERATORS.addAll 添加的都是操作符号。其中"+"、"-"、"*"、"/"、"%"、"div"、"&"、"|"、"^"、"~"符号大多用于 select 子句中。

/（除）运算符号运算的结果为 double 类型：

```
hive> select 10/3;
3.3333333333333335
```

div 操作符号也是除操作，运算的结果为整数类型：

```
hive> select 10 div 3;
3
```

%为取模操作，即计算余数

```
hive> select 10 % 3;
1
```

与（&）操作二进制运算，两个值必须都为 1 结果才是 1，否则为 0。下例中 2 的二进制为 10，1 二进制为 01，则 10 & 01 结果为 00，即为 0。

```
hive> select 2 & 1;
0
```

或（|）操作二进制运算，两个值只要有一个为 1，即为 1。2 的二进制为 10，1 的二进制为 01，则 10|01 的结果为 11，即结果为 3。

```
hive> select 2 | 1;
3
```

异或（^）操作二进制运算，只要两个值不一样，则为 1，一样时为 0。下例中 10^01 的结果为 11，即结果为 3。

```
hive> select 2^1;
3
```

按位取反（~）操作二进制运算，~1 为 0，~0 为 1。2 的二进制为 10。所以~10 的值为 1111111111111111111111111111101，即结果为-3。

```
hive> select  ~2;
-3
```

操作符号"=","==","<=>","!=","<>","<","<=",">",">="可用在 where 子句中，用于比较。
相等比较"=","==","<=>"具有相同的功能：

```
hive> select * from students where id==301;
hive> select * from students where id=="301";
hive> select * from students where id<=>"301";
```

操作符号"and","or","not","!"可用在 where 子句中，进行与、或、非运算：

```
hive> select * from students where id=301 and major='Oracle';
hive> select * from students where id=301 or major='Java';
```

8.8.2 更多函数

Hive 拥有丰富的内置函数。由于函数太多，以下仅为读者展示一部分。

（1）根据给定的元素，创建一个数组对象
语法：array(n0, n1, ...)

以下创建一个字符串数组对象：

```
hive> select array('Jack','Mary');
["Jack","Mary"]
```

以下创建一个整数的数组对象：

```
hive> select array(1,2,3);
[1,2,3]
```

下面例子由于包含一个字符串，所以创建一个字符串数组对象：

```
hive> select array(1,'Jack');
["1", "Jack"]
array_contains
```

（2）判断给定的元素是否存在数组中

语法：array_contains(array, value) - Returns TRUE if the array contains value

```
hive> select array_contains(array('Jack','Mary'),'Mary');
true
bin
```

（3）将一个 bigint 类型的数转成二进制

语法：bin(n) - returns n in binary

```
hive> select bin(2);
10
ceil
```

（4）返回大于当前数的最小整数

语法：ceil(x)

```
hive> select ceil(2.3);
3
current_date
```

（5）返回当前时间

语法：current_date()

```
hive> select current_date();
2021-3-18
current_timestamp
```

（6）返回当前的时间戳

语法：current_timestamp()

```
hive> select current_timestamp();
2021-3-18 21:37:44.271
explode
```

（7）将数组元素转换成多行显示

语法：explode(a)

```
hive> select  explode(array("Jack","Mary"));
Jack
Mary
```

也可以将一个 map 转换成多行显示：

```
hive> select  explode(map("name","Jack","age",34));
name    Jack
age     34
get_json_object
```

（8）根据指定的路径，解析出 json 字符串中的对象

json 中必须是""（双引号，即标准的 json 串），path 部分必须是以 "$." 开始。
如果只输入$，则表示当前整个 json 对象。

语法：get_json_object(json_txt, path)

```
hive> select get_json_object('{"name":"Jerry"}','$.name');
OK
Jerry
hive> select get_json_object('["Jack","Rose"]','$[1]');
OK
Rose
map
```

（9）创建一个 map 对象

语法：map(key0, value0, key1, value1,...)

```
hive> select map("name","Jack","age",34);
{"name":"Jack","age":"34"}
split
```

（10）根据指定的正则表达式将字符串转换成数组，正则表达式中要使用\\（两个斜线）

语法：split(str, regex)

```
hive> select split('Jack Mary Rose','\\s+');
["Jack","Mary","Rose"]
hive> select explode(split('Jack Mary Rose','\\s+'));
Jack
Mary
Rose
```

8.8.3 使用 Hive 函数实现 WordCount

首先使用 vim 创建一个文本文件：

```
$vim /hone/hadoop/notes.txt
```

里面的内容如下：

```
Hello this is a text for something
to tell you about how to process
wordcount in hive.
And you must be import into this file
into hive.
```

再创建一个 Hive 表，只包含一个列，且分割符号为'\r\n'，即回车换行。

```
hive>create table notes(text string)
>row format
>delimited fields terminated by '\r\n';
```

将 notes.txt 文件导入到 notes 表中：

```
hive> load data local inpath '/home/hadoop/notes.txt' into table notes;
```

测试查询是否是 5 行数据：

```
hive> select * from notes;
Hello this is a text for something
to tell you about how to process
wordcount in hive.
 And you must be import into this file
into hive.
Time taken: 0.159 seconds,
Fetched: 5 row(s)
```

再创建一个表，用于保存每一个单词：

```
hive> create table word(w string)
  > row format
  > delimited fields terminated by '\r\n';
```

现在我们需要将 notes 表中的每一行数据，按空格进行 split，然后再转换成行，保存到 word 表中去。以下使用 insert overwrite 语句，先将 word 表中的数据删除，然后再写入新的数据，如果使用 insert into 将会是追加数据。此语句会引发一个 MapReduce 计算。

```
hive> insert overwrite table word select explode(split(text,'\\s+')) from notes;
Total MapReduce CPU Time Spent: 2 seconds 740 msec
OK
Time taken: 22.725 seconds
```

先对 word 表进行查询，发现单词已经保存到 word 表中了：

```
hive> select * from word;
OK
this
is
a
text
for
something
to
tell
you
...
```

现在对 word 表进行 count 查询：

```
hive> select count(w),w from word group by w;
```

这个查询将会启动 MapReduce，并最终输出以下结果（部分内容略去）：

```
1    And
1    Hello
...
2    you
```

也可以将计算的结果保存到指定目录下：

```
hive> insert overwrite directory '/out001/' select concat(w,'\t',count(w)) from word group by w ;
```

直接在 Hive 里面查询数据:

```
hive> dfs -cat /out001/*;
And     1
a       1
about   1
be      1
file    1
for     1
hive.   2
...
```

8.9 本章小结

本章主要讲解了以下内容:

- Hive 是保存在 HDFS 上的数据库。Hive 是 HDFS 的一个客户端。
- schematool 命令用于初始化 Hive 的数据库,通过 dbType 指定数据库类型如 Derby 或 MySQL。
- 可以将 HDFS 数据映射成一张表,对这个表的操作就是对 HDFS 数据文件的分析。
- Hive 的数据对应的是 HDFS 的文件,Hive 的数据库对应的是 HDFS 的目录。
- Hive 执行的语句被称为 HQL,即 Hive Query Language。
- load data local inpath 用于导入 Linux 文件系统中的文件到 Hive 表中去。
- load data inpath 用于将 HDFS 上的文件移动到 Hive 表所在的目录下。
- Hive 默认使用 Deby 数据库作为 matedata。可以通过配置将 matedata 修改成 MySQL,这样就支持多用户同时使用同一个 matedata 了。
- Hive 拥有丰富的内置函数。可以使用 show functions 查看所有函数,使用"desc function 函数名;"查看某个具体函数的用法。

第 9 章

HBase 数据库

本章主要内容：

- HBase 的特点。
- HBase 的存储结构。
- HBase 操作命令。
- HBase 伪分布式。
- HBase 分布式。

HBase 是 Hadoop DataBase 的意思。HBase 是一种构建在 HDFS 之上的分布式、面向列的存储系统。在需要实时读写、随机访问超大规模数据集时，可以使用 HBase。

HBase 是 Google Bigtable 的开源实现，与 Google Bigtable 利用 GFS 作为其文件存储系统类似，HBase 利用 Hadoop HDFS 作为其文件存储系统，也是利用 HDFS 实现分布式存储的。Google 运行 MapReduce 来处理 Bigtable 中的海量数据，HBase 同样利用 Hadoop MapReduce 来处理 HBase 中的海量数据，Google Bigtable 利用 Chubby 作为协同服务，HBase 利用 ZooKeeper 作为协同服务。

9.1 HBase 的特点

HBase 具有如下特点：

- 大：一个表可以有上亿行，上百万列。
- 面向列：面向列表（族）的存储和权限控制，列（族）独立检索。
- 稀疏：对于为空（NULL）的列，并不占用存储空间，因此表可以设计得非常稀疏。
- 无模式：每一行都有一个可以排序的主键和任意多的列，列可以根据需要动态增加，同一张表中不同的行可以有截然不同的列。
- 数据多版本：每个单元中的数据可以有多个版本。默认情况下，版本号自动分配，版本号就是单元格插入时的时间戳。
- 数据类型单一：HBase 中的数据都是字符串，没有类型。

9.1.1 HBase 的高并发和实时处理数据

Hadoop 是一个高容错、高延时的分布式文件系统和高并发的批处理系统，不适用于提供实时计算；HBase 是可以提供实时计算的分布式数据库，数据被保存在 HDFS 分布式文件系统上，由 HDFS 保证其高容错性。那么在生产环境中，HBase 是如何基于 Hadoop 提供实时性呢？HBase 上的数据是以 StoreFile（HFile）二进制流的形式存储在 HDFS 上 block 块中；但是 HDFS 并不知道 HBase 存储的是什么，它只把存储文件看成为二进制文件，也就是说，HBase 的存储数据对于 HDFS 文件系统是透明的。

图 9-1 是 HBase 文件在 HDFS 上的存储示意图。

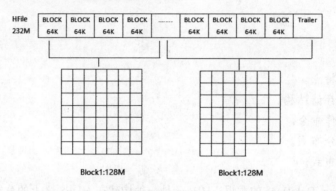

图 9-1

HBase HRegion Servers 集群中的所有的 region 数据，在服务器启动时都是被打开的，并且在内存初始化一些 memstore，相应地在一定程度上加快系统响应。而 Hadoop 的 block 中数据文件默认是关闭的，只有在需要的时候才打开，处理完数据后就关闭，这在一定程度上增加了响应时间。

9.1.2 HBase 的数据模型

HBase 以表的形式存储数据，表由行和列组成，列划分为若干个列族（row family），如图 9-2 所示。

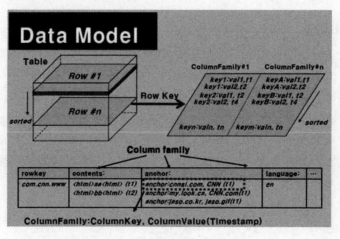

图 9-2

HBase 的逻辑数据模型如表 9-1 所示。

表 9-1 HBase 的逻辑数据模型

Row Key	Time Stamp	Column:"info"	Column:"other"		column:"..."
"key001"	t9		"other:name"	"Jerry"	
	t8		"other:age"	"100"	
	t6	"<HTML>..."			"col:value"
	t5	"Text ..."			
	t3	"Other..."			

1. 行键（Row Key）

与 NoSQL 数据库一样，Row Key 是用来检索记录的主键。访问 HBase table 中的行只有三种方式：

（1）通过单个 Row Key 访问。

（2）通过 Row Key 的 range 全表扫描。

（3）Row Key 可以使用任意字符串（最大长度为 64KB，实际应用中长度一般为 10～100bytes），在 HBase 内部，Row Key 保存为字节数组。

在存储时，数据按照 Row Key 的字典序（byte order）排序存储。设计 Key 时，要充分排序存储这个特性，将经常一起读取的行存储到一起（位置相关性）。注意：字典序对 int 排序的结果是 1,10,100,11,12,13,14,15,16,17,18,19,20,21,…,9,91,92,93,94,95,96,97,98,99。要保存整形的自然序，Row Key 必须用 0 进行左填充。

行的一次读写是原子操作（不论一次读写多少列）。这个设计决策能够使用户很容易理解程序在对同一个行进行并发更新操作时的行为。

2. 列族（Column Family）

HBase 表中的每个列都归属于某个列族。列族是表的 Schema 的一部分（而列不是），必须在使用表之前定义。列名都以列族作为前缀，例如 courses:history、courses:math 都属于 courses 这个列族。

3. 时间戳（Time Stamp）

HBase 中通过 Row 和 Columns 确定的一个存储单元称为 Cell。每个 Cell 都保存着同一份数据的多个版本。版本通过时间戳来索引，时间戳的类型是 64 位整型。时间戳可以由 HBase（在数据写入时自动）赋值，此时时间戳是精确到毫秒的当前系统时间。时间戳也可以由客户显示赋值。如果应用程序要避免数据版本冲突，就必须自己生成具有唯一性的时间戳。在每个 Cell 中，不同版本的数据按照时间倒序排序，即最新的数据排在最前面。

为了避免数据存在过多版本而造成的管理（包括存储和索引）负担，HBase 提供了两种数据版本回收方式。一是保存数据的最后 n 个版本，二是保存最近一段时间内的版本（比如最近七天）。用户可以针对每个列族进行设置。

4. Cell

Cell 是由{row key，column(=< family> + < label>)，version}唯一确定的单元。Cell 中的数据是没有类型的，全部都以字节码的形式存储。

9.2　HBase 的安装

HBase 可能独立运行在一台主机上。此时 HBase 将只会有一个进程，即 HMaster（这个 HMaster 内含 HRegionServer 和 HQuorumPeer 两个服务）。虽然只有一个进程，但也可以提供 HBase 的大部分功能。

首先需要了解 HBase 的版本信息，在笔者写作本书时，HBase 的最高版本为 2.4，我们可以在 HBase 的官方网站上，通过 hbase documents 查看 HBase 对 JDK、Hadoop 版本的兼容性。首先查看 HBase 对 JDK 版本的支持如表 9-2 所示。

表 9-2　HBase 对 JDK 版本的支持

Java Version	HBase 1.4+	HBase 2.2+	HBase 2.3+
JDK7	√	×	×
JDK8	√	√	√
JDK11	×	×	*

表中的×表示不支持，√表示支持，*表示未测试。通过上表可以看出，任何版本都支持 JDK1.8+。而我们之前安装的环境，也正是基于 JDK1.8 的。

我们再来看一下，HBase 对 Hadoop 版本的支持情况，如表 9-3 所示。

表9-3　对Hadoop版本的支持

	HBase-1.4.x	HBase-1.6.x	HBase-1.7.x	HBase-2.2.x	HBase-2.3.x
Hadoop-2.7.0	✗	✗	✗	✗	✗
Hadoop-2.7.1+	✓	✗	✗	✗	✗
Hadoop-2.8.[0-2]	✗	✗	✗	✗	✗
Hadoop-2.8.[3-4]	!	✗	✗	✗	✗
Hadoop-2.8.5+	!	✓	✓	✓	✗
Hadoop-2.9.[0-1]	✗	✗	✗	✗	✗
Hadoop-2.9.2+	!	✓	✓	✓	✗
Hadoop-2.10.x	!	✓	✓	!	✓
Hadoop-3.1.0	✗	✗	✗	✗	✗
Hadoop-3.1.1+	✗	✗	✗	✓	✓
Hadoop-3.2.x	✗	✗	✗	✓	✓

基于上面的配置，本次示例将选择 HBase2.3。所以现在就动手下载 HBase，下载的地址为 https://mirrors.tuna.tsinghua.edu.cn/apache/hbase/2.3.4/hbase-2.3.4-bin.tar.gz。

9.2.1　HBase 的单节点安装

HBase 单节点安装可以让我们快速学会 HBase 的基本使用。HBase 的单节点安装不需要 Hadoop，数据保存到指定的磁盘目录下。在 hbase-site.xml 文件中，通过 "hbase.rootdir=file:///" 来指定数据保存的目录。也不需要 ZooKeeper，启动后 HBase 将使用自带的 ZooKeeper。只有一个进程 Hmaster（内部包含：HRegionServer 和 HQuorumPeerman 两个子线程）。以下是 HBase 单节点安装的过程。

步骤 01 上传并解压 HBase。

同样地，使用 tar 将 HBase 解压到 /app/ 目录下：

```
$ tar -zxvf hbase-2.3.4-bin.tar.gz -C /app/
```

步骤 02 修改配置文件。

首先修改的是 hbase-env.sh 文件，此文件中保存了 JAVA_HOME 环境变量信息。配置如下：

```
$ vim /app/hbase-2.3.4/conf/hbase-env.sh
export JAVA_HOME=/usr/java/jdk1.8.0_281
export HBASE_MANAGES_ZK=true
```

再修改 hbase-site.xml 配置文件，此配置文件中，我们需要指定 HBase 数据的保存目录，由于我们的目的是快速入门，所以可以先将 HBase 的数据保存到磁盘上。具体的配置如下：

```
$ vim /app/hbase-2.3.4/conf/hbase-site.xml
<configuration>
    <property>
        <name>hbase.rootdir</name>
        <value>file:///app/datas/hbase</value>
    </property>
    <property>
        <name>hbase.zookeeper.property.dataDir</name>
        <value>/app/datas/zookeeper</value>
    </property>
    <!--以下配置是否检查流功能 -- >
    <property>
        <name>hbase.unsafe.stream.capability.enforce</name>
        <value>false</value>
    </property>
</configuration>
```

修改 regionservers 文件，此文件用于指定 HRegionServer 的服务器地址，由于是单机部署，所以指定本机名称即可。添加本机主机名：

```
$ vim regionservers
server201
```

步骤 03 启动/停止 HBase。

启动 HBase，只需要在 HBase 的 bin 目录下，执行 start-hbase.sh 即可。

```
[hadoop@server201 app]$ hbase-2.3.4/bin/start-hbase.sh
SLF4J: Class path contains multiple SLF4J bindings.
SLF4J: Actual binding is of type [org.slf4j.impl.Log4jLoggerFactory]
```

启动完成以后，通过 jps 查看进程，会发现 HMaster 进程已经运行：

```
[hadoop@server201 app]$ jps
7940 Jps
7785 HMaster
```

停止 HBase 只需要执行 stop-hbase.sh 即可：

```
[hadoop@server201 app]$ hbase-2.3.4/bin/stop-hbase.sh
stopping hbase...
```

步骤 04 登录 HBase Shell。

在 bin 目录下，使用 hbase shell 命令登录 HBase Shell，即可操作 HBase 数据库。登录成功后，将显示 hbase >命令行。

```
[hadoop@server201 app]$ hbase-2.3.4/bin/hbase shell
HBase Shell
Use "help" to get list of supported commands.
Use "exit" to quit this interactive shell.
For Reference, please visit: http://hbase.apache.org/2.0/book.html#shell
Version 2.3.4, rafd5e4fc3cd259257229df3422f2857ed35da4cc, Thu Jan 14 21:32:25 UTC 2021
Took 0.0006 seconds
hbase(main):001:0>
```

步骤 05 HBase 数据操作。

现在，我们快速创建一个表，保存一些数据，以了解 HBase 是如何保存数据的。首先，不了解 HBase 命令的读者，可以直接在 HBase 的命令行输入 help，此命令将会显示 HBase 的帮助信息。

创建一个命名空间，可以理解成创建了一个数据库：

```
hbase(main):009:0> create_namespace 'ns1'
Took 0.1776 seconds
```

查看所有命名空间，可以理解成查看所有数据库：

```
hbase(main):010:0> list_namespace
NAMESPACE
default
hbase
ns1
3 row(s)
Took 0.0158 seconds
```

创建一个表，在指定的命名空间下，并指定列族为"f"。可以理解为创建一个表，并指定一个列名为 f：

```
hbase(main):011:0> create 'ns1:stud','f'
Created table ns1:stud
Took 0.8601 seconds
=> Hbase::Table - ns1:stud
```

向表中写入一行记录，其中 R001 为主键，即 Row Key，f:name 为列名，Jack 为列值。此处与关系型数据库有很大的区别，注意区分。

```
hbase(main):012:0> put 'ns1:stud','R001','f:name','Jack'
Took 0.3139 seconds
hbase(main):013:0> put 'ns1:stud','R002','f:age','34'
Took 0.0376 seconds
```

查询表，类似于关系型数据库中的 select：

```
hbase(main):014:0> scan 'ns1:stud'
ROW                     COLUMN+CELL
R001                    column=f:name, timestamp=1568092417316, value=Jack
R002                    column=f:age, timestamp=1568092435076, value=34
2 row(s)
Took 0.0729 seconds
```

9.2.2 HBase 的伪分布式安装

HBase 的伪分布式安装，是将 HBase 的数据保存到伪分布式的 HDFS 系统中。此时 Hadoop 环境为伪分布式，HBase 的节点只有一个，HBase 使用独立运行的 ZooKeeper。以下步骤将配置一个 HBase 的伪分布式运行环境。我们需要：

- 配置好的 Hadoop 运行环境。
- 独立运行的 ZooKeeper，只有一个节点即可。
- 将 Hadoop 的 ZooKeeper 指向这个独立的 ZooKeeper。

步骤 01 准备 Hadoop 和 ZooKeeper。

请根据前面相关章节的步骤，配置好 Hadoop 的伪分布式运行环境。
请根据前面相关章节的步骤，配置好 ZooKeeper 的独立节点运行环境。
测试以上两个环境正常运行并可用。

步骤 02 修改 HBase 配置文件。

首先修改 HBase 的配置文件 hbase-env.sh 文件，在此配置文件中，重点配置 ZooKeeper 选项，配置为使用外部的 ZooKeeper 即可。具体配置如下：

```
$ vim /app/hbase-2.3.4/conf/hbase-env.sh
export JAVA_HOME=/usr/java/jdk1.8.0_281
export HBASE_MANAGES_ZK=false
```

修改 HBase 配置文件 hbase-site.xml 文件，重点关注 hbase.rootdir 用于指定在 Hadoop 中存储 HBase 数据的目录。ZooKeeper 用于指定 ZooKeeper 地址。

```xml
<configuration>
    <property>
        <name>hbase.cluster.distributed</name>
        <value>true</value>
    </property>
    <property>
        <name>hbase.tmp.dir</name>
        <value>/app/datas/hbase/tmp</value>
    </property>
    <property>
        <name>hbase.unsafe.stream.capability.enforce</name>
        <value>false</value>
    </property>
    <property>
        <name>hbase.rootdir</name>
        <value>hdfs://server201:8020/hbase</value>
    </property>
    <property>
        <name>hbase.zookeeper.quorum</name>
        <value>server201:2181</value>
    </property>
</configuration>
```

步骤 03 启动 HBase。

直接使用 start-hbase.sh 启动 HBase：

```
$ /app/hbase-2.3.4/bin/start-hbase.sh
```

查看 HBase 的进程如下：

```
[hadoop@server201 conf]$ jps
2992 NameNode
3328 SecondaryNameNode
5793 Jps
5186 Main
3560 ResourceManager
4120 QuorumPeerMain
4728 HRegionServer
3691 NodeManager
4523 HMaster
3119 DataNode
```

同样地，可以通过浏览器查看 16010 端口，如图 9-4 所示。

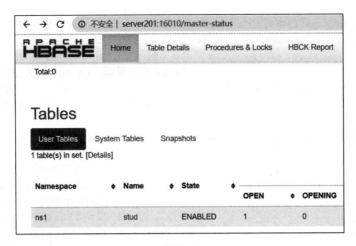

图 9-4

9.2.3 Java 客户端代码

在 HBase 的分布式启动成功后，就可以通过 Java 代码去操作 HBase 中的数据。本小节将讲解部分操作代码。

使用 Java 代码操作需要添加两个依赖，一个是 Hadoop 的，因为 HBase 依赖的某些配置对象是 Hadoop 包中的；一个是依赖 hbase-client。具体依赖如下：

```xml
<dependency>
    <groupId>org.apache.hbase</groupId>
    <artifactId>hbase-client</artifactId>
    <version>2.3.4</version>
</dependency>
<dependency>
    <groupId>org.apache.hadoop</groupId>
    <artifactId>hadoop-client</artifactId>
    <version>3.2.2</version>
</dependency>
```

【代码 9-1】ConnectTest.java 测试与 HBase 的连接（注意，我们连接的端口为 ZooKeeper 的端口 2181）

```
1.  package org.hadoop.hbase;
2.  @Slf4j
3.  public class ConnectTest {
4.      public static void main(String[] args) throws Exception {
5.          Configuration configuration = HBaseConfiguration.create();
6.          configuration.set("hbase.zookeeper.property.clientPort", "2181");
7.          configuration.set("hbase.zookeeper.quorum", "server201");
8.          Connection connection = ConnectionFactory.createConnection(configuration);
9.          log.info("测试连接："+connection);
10.         Admin admin = connection.getAdmin();
11.         NamespaceDescriptor[] ns = admin.listNamespaceDescriptors();
```

```
12.          log.info("命名空间个数："+ns.length);
13.          for(NamespaceDescriptor n:ns){
14.              System.out.println("命名空间："+n.getName());
15.          }
16.          connection.close();
17.     }
18. }
```

测试输出结果：

测试连接：hconnection-0x131774fe
命名空间个数：3
命名空间：default
命名空间：hbase
命名空间：ns1

为了给读者展示更多的示例，下面我们通过一个 Junit 测试来操作 HBase，这些操作中包含查询表名称、查询表中的数据等。

【代码 9-2】HBaseOperation.java

```
1.  package org.hadoop.hbase;
2.  @Slf4j
3.  public class HBaseOperation {
4.      private Connection con;
5.      @Before
6.      public void before() throws Exception {
7.          Configuration configuration = HBaseConfiguration.create();
8.          configuration.set("hbase.zookeeper.property.clientPort", "2181");
9.          configuration.set("hbase.zookeeper.quorum", "server201");
10.         con = ConnectionFactory.createConnection(configuration);
11.     }
12.     @After
13.     public void after() throws Exception {
14.         con.close();
15.     }
16.     /**
17.      * 查询所有表
18.      */
19.     @Test
20.     public void listTables() throws Exception {
21.         Admin admin = con.getAdmin();
22.         List<TableDescriptor> list = admin.listTableDescriptors();
23.         for (TableDescriptor t : list) {
24.             log.info("tableName:" + t.getTableName().getNameAsString());
25.         }
26.     }
27.     /**
28.      * 查询指定命名空间的表
29.      */
30.     @Test
31.     public void listNamespaceTables() throws Exception {
```

```
32.        Admin admin = con.getAdmin();
33.        TableName[] tns = admin.listTableNamesByNamespace("ns1");
34.        for (TableName tn : tns) {
35.            log.info("name :" + tn.getNameAsString());
36.        }
37.        con.close();
38.    }
39.    /**
40.     * 查询表中的数据
41.     */
42.    @Test
43.    public void queryData() throws Exception {
44.        Table table = con.getTable(TableName.valueOf("ns1:stud"));
45.        Scan scan1 = new Scan();
46.        ResultScanner rs = table.getScanner(scan1);
47.        for (Result r : rs) {
48.            for (Cell cell : r.listCells()) {
49.                log.info("RowKey:" + Bytes.toString(CellUtil.cloneRow(cell)) + "\t"
50.                        + Bytes.toString(CellUtil.cloneFamily(cell)) + "\t"
51.                        + Bytes.toString(CellUtil.cloneQualifier(cell)) + "\t"
52.                        + Bytes.toString(CellUtil.cloneValue(cell)));
53.            }
54.            log.info("--------------");
55.        }
56.        rs.close();
57.    }
58.    /**
59.     * 创建一个表
60.     */
61.    @Test
62.    public void createTable() throws Exception {
63.        Admin admin = con.getAdmin();
64.        TableDescriptor td =
65.                TableDescriptorBuilder
66.                        .newBuilder(TableName.valueOf("ns1:person"))//表名
67.                        .setColumnFamily(ColumnFamilyDescriptorBuilder.of("c"))//列族
68.                        .build();//创建 TableDescriptor
69.        admin.createTable(td);//创建 Table
70.    }
71.    /**
72.     * 删除一个表
73.     */
74.    @Test
75.    public void dropTable() throws Exception {
76.        Admin admin = con.getAdmin();
77.        admin.disableTable(TableName.valueOf("ns1:person"));
78.        admin.deleteTable(TableName.valueOf("ns1:person"));
79.        admin.close();
```

```
80.     }
81.     /**
82.      * 向表中插入数据(如果 key 值已经存在，则为修改)
83.      */
84.     @Test
85.     public void putData() throws Exception {
86.         Table table = con.getTable(TableName.valueOf("ns1:stud"));
87.         Put put = new Put("K002".getBytes());
88.         put.addColumn("f".getBytes(), "name".getBytes(), "Rose".getBytes());
89.         table.put(put);
90.         table.close();
91.     }
92.     /**
93.      * 根据 rowke 删除记录
94.      */
95.     @Test
96.     public void deleteByRowKey() throws Exception {
97.         Table table = con.getTable(TableName.valueOf("ns1:stud"));
98.         Delete delete = new Delete(Bytes.toBytes("K002"));
99.         table.delete(delete);
100.        table.close();
101.    }
102.    /**
103.     * 删除一个单元格，则需要在 Delete 中输入一个 Column 即可
104.     */
105.    @Test
106.    public void deleteColumn() throws Exception {
107.        String key = "K001";
108.        Table table = con.getTable(TableName.valueOf("ns1:stud"));
109.        boolean boo = table.exists(new Get(Bytes.toBytes(key)));
110.        if (boo) {
111.            Delete delete = new Delete(Bytes.toBytes(key));
112.            //设置需要删除的列即可以只删除一个单元格
113.            delete.addColumn(Bytes.toBytes("f"), Bytes.toBytes("name"));
114.            table.delete(delete);
115.        } else {
116.            log.info("此 rowkey 不存在:" + key);
117.        }
118.        table.close();
119.    }
120. }
```

9.2.4 其他 Java 操作代码

其他更多的 Java 操作代码不胜枚举，下面直接列出这些代码。

（1）通过 Row Key 查询

只要在 Scan 中传递 Get 对象即可是 Row Key 查询。比如：

```
Scan scan = new Scan(new Get("U001".getBytes()));
```

(2) 根据值进行查询

根据值进行查询，可以使用 ValueFilter 对象，BinaryComparator 用于比较二进制数据。比如：

```
Scan scan = new Scan();
//设置查询列的信息
BinaryComparator bc = new BinaryComparator("Mary".getBytes());
ValueFilter vf = new ValueFilter(CompareOp.EQUAL, bc);
scan.setFilter(vf);
```

(3) 使用正则表达式查询

RegexStringComparator 用于执行正则表达式的查询。比如：

```
RegexStringComparator bc = new RegexStringComparator(".*r.*");
ValueFilter vf = new ValueFilter(CompareOp.EQUAL, bc);
scan.setFilter(vf);
```

或使用 SingleColumnValueFilter 指定查询的列名：

```
RegexStringComparator bc = new RegexStringComparator(".*r.*");
SingleColumnValueFilter vf =
    new SingleColumnValueFilter(
    Bytes.toBytes("info"),
    Bytes.toBytes("name"),
    CompareOp.EQUAL , bc);
scan.setFilter(vf);
```

(4) 字符串包含查询

可以使用 SubStringComparator 查询包含的字符串。比如：

```
ByteArrayComparable bc =new SubstringComparator("Mary");
```

(5) 前缀二进制比较器

BinaryPrefixComparator 是前缀二进制比较器。与二进制比较器不同的是，它只比较前缀是否相同。比如：以下查询 info:name 列以"Ma"为前缀的数据。

```
Scan scan = new Scan();
BinaryPrefixComparator comp
= new BinaryPrefixComparator(Bytes.toBytes("Ma"));
SingleColumnValueFilter filter
= new SingleColumnValueFilter(Bytes.toBytes("info"),
Bytes.toBytes("name"), CompareOp.EQUAL, comp);
scan.setFilter(filter);
```

(6) 列值过滤器

SingleColumnValueFilter 用于测试值的情况为相等、不等、范围，等等。

下面代码检测列族 family 下的列 qualifier 的值和字符串"some-value"相等的部分：

```
Scan scan = new Scan();
SingleColumnValueFilter filter
= new SingleColumnValueFilter(Bytes.toBytes("family"),
```

```
            Bytes.toBytes("qualifier"),
CompareOp.EQUAL, Bytes.toBytes("some-value"));
scan.setFilter(filter);
```

(7)排除过滤

SingleColumnValueExcludeFilter 与 SingleColumnValueFilter 功能一样，只是不查询出该列的值。下面代码就不会查询出 family 列族下 qualifier 列的值：

```
Scan scan = new Scan();
SingleColumnValueExcludeFilter filter
= new SingleColumnValueExcludeFilter(Bytes.toBytes("family"),
            Bytes.toBytes("qualifier"),
CompareOp.EQUAL, Bytes.toBytes("some-value"));
scan.setFilter(filter);
```

(8)列族过滤器

FamilyFilter 用于过滤列族（通常在 Scan 过程中通过设定某些列族来实现该功能，而不是直接使用该过滤器）。比如：

```
Scan scan = new Scan();
FamilyFilter filter
= new FamilyFilter(CompareOp.EQUAL,
new BinaryComparator(Bytes.toBytes("some-family")));
scan.setFilter(filter);
```

(9)列名过滤器

QualifierFilter 用于列名（Qualifier）过滤。比如：

```
QualifierFilter qff = new QualifierFilter(CompareOp.EQUAL,
            new BinaryComparator("name".getBytes()));
scan.setFilter(qff);
```

(10)列名前缀过滤器

ColumnPrefixFilter 用于列名（Qualifier）前缀过滤，即包含某个前缀的所有列名。比如：

```
Scan scan = new Scan();
ColumnPrefixFilter filter =
    new ColumnPrefixFilter(Bytes.toBytes("somePrefix"));
scan.setFilter(filter);
```

(11)多个列名前缀过滤器

MultipleColumnPrefixFilter 与 ColumnPrefixFilter 的行为类似，但可以指定多个列名（Qualifier）前缀。比如：

```
Scan scan = new Scan();
byte[][] prefixes = new byte[][]{
Bytes.toBytes("prefix1"),
        Bytes.toBytes("prefix2")};
MultipleColumnPrefixFilter filter =
new MultipleColumnPrefixFilter(prefixes); scan.setFilter(filter);
```

（12）列范围过滤器

ColumnRangeFilter 该过滤器可以进行高效的列名内部扫描。比如：

```
Scan scan = new Scan();
boolean minColumnInclusive = true;
boolean maxColumnInclusive = true;
ColumnRangeFilter filter =
new ColumnRangeFilter(
Bytes.toBytes("minColumnName"), minColumnInclusive,
Bytes.toBytes("maxColumnName"), maxColumnInclusive);
scan.setFilter(filter);
```

（13）行键过滤器

RowFilter 行键过滤器，一般来讲，执行 Scan 使用 startRow/stopRow 方式比较好，而 RowFilter 过滤器也可以完成对某一行的过滤。比如：

```
Scan scan = new Scan();
RowFilter filter =
new RowFilter(CompareOp.EQUAL,
new BinaryComparator(Bytes.toBytes("someRowKey1")));
scan.setFilter(filter);
```

（14）分页过滤器

PageFilter 用于按行分页。必须设置每次显示几行，以及在 Scan 中设置开始的行值。比如：

```
Scan scan = new Scan();
PageFilter pf = new PageFilter(5);
scan.setFilter(pf);
byte[] startRow
 = Bytes.add("U005".getBytes(),Bytes.toBytes("postfix"));
scan.setStartRow(startRow);
ResultScanner resultScanner = table.getScanner(scan);
```

（15）串联多个过滤器

可以使用 FilterList 将多个过滤器串联起来，组成 And 或 Or 的过滤器。比如：

```
FilterList fl = new FilterList(Operator.MUST_PASS_ALL);         fl.addFilter(new
ValueFilter(CompareOp.EQUAL,
new BinaryComparator("Smith".getBytes())));
fl.addFilter(new RowFilter(CompareOp.EQUAL,
new BinaryComparator("U005".getBytes())));
scan.setFilter(fl);
```

由于过滤器比较多，更多的过滤器用法，请读者查阅官方文档，这里不再一一展示。

9.3　HBase 集群安装

在 HBase 集群的情况下，会有一个 HMaster 进程和多个 HRegionServer 进程，其中 HRegionServer

会根据配置的节点情况运行在多台主机上。以下将介绍如何安装一个 HBase 的集群环境。在安装时,要注意 HBase 与 Hadoop 的兼容关系。三台服务器的配置如表 9-4 所示。

表9-4 三台服务器的配置

Ip/name	软件	进程
192.168.56.101 server101	Hadoop-3.2.2 Zookeeper-3.6.2 HBase-2.3.4	NameNode SecondaryNameNode ResourceManager NodeManager DataNode QuorumPeerMan HMaster HRegionServer
192.168.56.102 server102	Hadoop-3.2.2 Zookeeper-3.6.2 HBase-2.3.4	DataNode NodeManager QuorumPeerMan HRegionServer
192.168.56.103 server103	Hadoop-3.2.2 Zookeeper-3.6.2 HBase-2.3.4	DataNode NodeManager QuorumPeerMan HRegionServer

安装 ZooKeeper 集群和 Hadoop 集群的过程请参考之前的章节。以下直接安装 HBase。

步骤 01 上传并解压 HBase 安装包。

```
$tar -zxvf ~/hbase-2.3.4-bin.tar.gz  -C  /app/
```

目录名称太长,修改一下名称:

```
$ mv /app/hbase-2.3.4-bin /app/hbase-2.3.4
```

步骤 02 配置环境变量。

添加 HBase 环境变量:

```
$ sudo vim /etc/profile
export HBASE_HOME=/app/hbase-2.3.4
export PATH=$PATH:$HBASE_HOME/bin
```

让环境变量生效:

```
$ source /etc/profile
```

步骤 03 配置 HBase。

修改 hbase-env.sh 文件:

```
export JAVA_HOME=/usr/java/jdk1.8.0_281
export HBASE_MANAGES_ZK=false
```

配置 hbase-site.sh 文件:

```xml
<configuration>
    <property>
        <name>hbase.tmp.dir</name>
        <value>/app/datas/hbase/tmp</value>
    </property>
    <property>
        <name>hbase.rootdir</name>
        <value>hdfs://server201:8020/hbase</value>
    </property>
    <property>
        <name>hbase.cluster.distributed</name>
        <value>true</value>
    </property>
    <property>
        <name>hbase.zookeeper.quorum</name>
        <value>server101:2181,server102:2181,server103:2181</value>
    </property>
    <property>
        <name>hbase.unsafe.stream.capability.enforce</name>
        <value>false</value>
    </property>
</configuration>
```

- hbase.rootdir：用于配置 hfs，即 HBase 的文件系统所保存的目录。
- hbase.tmp.dir：用于配置 HBase 临时文件的保存目录，如果不设置，临时文件将保存到/tmp 目录下。
- hbase.zookeeper.property.dataDir：用于配置 HBase 自带的 ZooKeeper 数据文件保存的目录。
- hbase.zookeeper.quorum：用于设置外部 ZooKeeper 的地址和目录。需要在 hbase-env.sh 中将 ZK 设置为 false。

配置 regionservers：

```
server101
server102
server103
```

将所有关于 HBase 的配置分发到其他主机，包含环境变量配置文件和 HBase 整个目录：

```
$ scp /etc/profile root@server102:/etc/
$ scp /etc/profile root@server103:/etc/
$ scp -r /app/hbase-2.3.4 server102:/app/
$ scp -r /app/hbase-2.3.4 server103:/app/
```

步骤 04 启动 HBase。

与前面一样，执行$HBASE_HOME/bin 目录下的 start-hbase.sh 即可启动 HBase。由于是集群配置，需要先启动 Hadoop 和 ZooKeeper。执行 start-hbase.sh，脚本将会根据配置启动其他节点上的 HRegionServer。

```
$ /app/hbase-2.3.4/bin/start-hbase.sh
```

jps 命令查看进程：

```
[hadoop@server101 app]$ jps
9073 ResourceManager
22433 HRegionServer
5442 DataNode
23075 Jps
9205 NodeManager
5676 SecondaryNameNode
17357 QuorumPeerMain
5310 NameNode
22270 HMaster
```

查看其他节点的进程，如 server102 主机上的进程：

```
[hadoop@server102 app]$ jps
22433 HRegionServer
5446 DataNode
23015 Jps
9203 NodeManager
17337 QuorumPeerMain
```

步骤 05 访问 16010 端口。

使用浏览器访问 16010 端口，将会看到 HBase 整个集群的情况。如图 9-5 所示。

图 9-5

如果要启动高可靠，则可以在 server102（即另一台机器）上再启动一个 master，使用以下命令：

```
$ /app/hbase-2.3.4/bin/hbase-deamon.sh start master
```

启动以后，访问 server102 上的 16010 端口，此时 server102 会显示为 Backup master，即为备份节点，如图 9-6 所示。

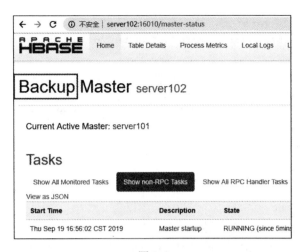

图 9-6

注意：启动 HBase 之前，必须确定已经在/etc/hosts 文件中配置了本地 DNS。

使用 http://ip:16010 访问之前，必须确定防火墙已经关闭。关闭防火墙使用$sudo systemctl stop firewalld.service。禁用防火墙使用$sudo systemctl disable firewalld.service。

9.4 HBase Shell 操作

HBase Shell 提供了大多数的 HBase 命令，通过 HBase Shell 用户可以方便地创建、删除及修改表，还可以向表中添加数据、列出表中的相关信息等。在启动 HBase 之后，用户可以通过下面的命令进入 HBase Shell 命令行模式。HBase 脚本文件在$HBASE_HOME/bin/目录下。

```
$./hbase shell
```

在登录成功 hbase shell 以后，可以使用 help 显示所有命令列表：

```
hbase(main):002:0> help
```

输入 help 可以查看到命令分组，如表 9-5 所示（注意，命令都是小写，有大小写的区分）。

表 9-5 HBase 命令分组

命令分组	命 令
general	status、version
ddl	alter、create、describe、disable、drop、enable、exists、is_disable、is_enable、list
dml	count、delete、deleteall、get_counter、incr、put、scan、truncate
tools	assign、blance_switch、blancer、close_region、compact、flush、major_compact、move、split、unassign...
replication	add_peer、append_peer_tableCFs、disable_peer、disable_table_replication、enable_peer....
...	...

由于命令比较多，请读者自行使用 help 查看所有命令。以下将演示一些简单命令的基本操作。

查看版本信息 version：

```
hbase(main):008:0> version
2.3.4,r930b9a55528fe45d8edce7af42fef2d35e77677a, Thu Apr  6 19:36:54 PDT 2021
```

查看状态 status：

```
hbase(main):009:0> status
1 active master, 0 backup masters, 1 servers, 1 dead, 2.0000 average load
```

9.4.1 DDL 操作

还有一些 DDL 操作，如 create、alter、desc 等。

（1）创建一个表 create

语法 1：create "表名","列族 1","列族 2","列族 N"
语法 2：create "表名",{NAME=>"列族 1",VERSIONS=>保存版本数量},{...}
保存版本数量，用于记录一个列族最多可以保存的历史记录。

```
hbase(main):005:0> create "stud","info"
0 row(s) in 1.4040 seconds
=> Hbase::Table - stud
hbase(main):006:0> list
TABLE
stud
1 row(s) in 0.0290 seconds
=> ["stud"]
hbase(main):007:0> describe "stud"
```

上面的代码中，stud 为表名，可以使用""（双引号），也可以使用''（单引号）。info 为列族。或者指定更多的信息，如使用 VERSIONS 指定保存的版本信息：

```
hbase(main):008:0> create "person",{NAME=>"info",VERSIONS=>3}
0 row(s) in 1.2580 seconds
=> Hbase::Table - person
hbase(main):009:0> list
TABLE
person
stud
2 row(s) in 0.0250 seconds
=> ["person", "stud"]
```

person 为表名，在 person 后面通过 "{}" 声明列族为 info，版本信息为 3。创建表以后，通过 list 可以显示所有的数据表。

（2）修改表结构，添加一个新的列，可以使用 alter
语法：alter "表名",{NAME=>"列族名称",VERSIONS=>3}

```
hbase> alter 'stud', NAME => 'info', VERSIONS => 5
```

上面语句修改列族最大单元数。如果修改的列族不存在，则为添加一个新的列族，否则为修改已有的列族信息。在修改之前，可以先通过 desc 查看表的信息；在修改之后，可以通过 desc 查看表的信息来确认。

```
hbase(main):003:0> desc "stud"
```

（3）查看表的信息，会列出这个表的所有列族信息

```
hbase(main):003:0> desc "stud"
Table stud is ENABLED
stud
COLUMN FAMILIES DESCRIPTION
{NAME => 'info', BLOOMFILTER => 'ROW', VERSIONS => '1',...
```

通过上面的列族信息可以看出，info 列族的 versions 为 1，现在可以通过 alter 修改 info 列族的 versions 为 3：

```
hbase(main):004:0> alter "stud",{NAME=>"info",VERSIONS=>3}
```

然后，再通过 desc 查看 stud 表的信息：

```
hbase(main):005:0> desc "stud"
Table stud is ENABLED
stud
COLUMN FAMILIES DESCRIPTION
{NAME => 'info', BLOOMFILTER => 'ROW', VERSIONS => '3',...
```

此时可以看到，versions 已经修改为 3 了。也可以通过 alter 添加一个新的列族，如果这个列族不存在，则为创建一个新的列族：

```
hbase(main):006:0> alter "stud",{NAME=>"desc",VERSIONS=>3}
```

然后再通过 desc 查看 stud 表的结构：

```
hbase(main):007:0> desc "stud"
Table stud is ENABLED
stud
COLUMN FAMILIES DESCRIPTION
{NAME => 'desc', BLOOMFILTER => 'ROW', VERSIONS => '3',...
{NAME => 'info', BLOOMFILTER => 'ROW', VERSIONS => '3',...
```

可见，表中已经添加了一个新的列族 desc。

注意：HBase 的列族信息是按字典顺序来排序的，所以 desc 的列族信息在 info 列族信息的前面。

（4）删除一个列族

语法：alter "表名",{NAME=>"列族名称",METHOD=>"delete"}

删除一个列族，同样使用 alter 关键字，只是必须传递 method=>"delete" 来指定删除。

删除一个列族：

```
hbase(main):009:0> alter "stud",{NAME=>"desc",METHOD=>"delete"}
```

再次查看这个表的信息：

```
hbase(main):010:0> desc "stud"
{NAME => 'info', BLOOMFILTER => 'ROW', VERSIONS => '3',...
```

可见，表中只剩下一个列族 info 了，即 desc 列族已经被删除。

9.4.2 DML 操作

DML 操作用于向表中写入数据、查询数据及删除数据。功能类似于 SQL 语句的 DML，但命令与 SQL 不同。

（1）插入数据

语法：put "表名","行键","列族:列名","具体值"

HBase 写入的数据没有数据类型，都是二进制数据。相同的列族属于同一行数据。

向表中写入一行数据：

```
hbase(main):012:0> put "stud","U001","info:name","Jack"
```

上例中，stud 为表名，U001 为行键即 Row Key 的值，info 后面的 name 为列的名称。Jack 为列值。插入数据以后，就可以通过表描述显示表中的所有数据了。

（2）扫描表中的数据

语法：scan "表名"

scan 用于显示表中的所有数据，类似于 SQL 语法中的 select *from xxx 语句。

```
hbase(main):013:0> scan "stud"
ROW    COLUMN+CELL
U001   column=info:name, timestamp=1500782803718, value=Jack
1 row(s) in 0.0710 seconds
```

上例中最后一行 1 row(s)…说明目前只有一行数据。现在就可以写入多行数据，然后再通过 scan 查看里面的数据了。

再写入一行数据，行键即 Row Key 与前面的行键相同：

```
hbase(main):014:0> put "stud","U001","info:age",23
```

然后再扫描表中的记录，依然是一行数据，因为一个行键代表的是一行数据。

```
hbase(main):015:0> scan "stud"
ROW    COLUMN+CELL
U001   column=info:age, timestamp=1500783145216, value=23
U001   column=info:name, timestamp=1500782803718, value=Jack
row(s) in 0.0280 seconds
```

现在写入一个不同的行键，然后再进行查询测试：

```
hbase(main):016:0> put "stud","U002","info:name","Mary"
```

然后再查看表中的记录，已经发现表中存在两行记录了，因为行键即 Row Key 不同。

```
hbase(main):017:0> scan "stud"
ROW    COLUMN+CELL
U001   column=info:age, timestamp=1500783145216, value=23
U001   column=info:name, timestamp=1500782803718, value=Jack
U002   column=info:name, timestamp=1500783317379, value=Mary
2 row(s) in 0.0280 seconds
```

可以通过 put 多次修改列的值，首先查看 U001 和 info:name 的值：

```
hbase(main):018:0> scan "stud"
ROW     COLUMN+CELL
U001      column=info:name, timestamp=1500782803718, value=Jack
...
2 row(s) in 0.0520 seconds
```

上面 info:name 的值为 Jack，以下分别修改 N 次：

```
hbase(main):019:0> put "stud","U001","info:name","FirstName"
hbase(main):020:0> put "stud","U001","info:name","SecondName"
hbase(main):021:0> put "stud","U001","info:name","ThirdName"
```

然后再扫描表，info:name 值为最后一次修改的记录：

```
hbase(main):022:0> scan "stud"
ROW     COLUMN+CELL
U001      column=info:name, timestamp=1500783878309, value=ThirdName
...
```

HBase 的 versions 主要控制数据的修改操作版本，上面对列 info:name 的数据修改了 3 次，可以通过版本扫描显示所有修改过的记录。

（3）扫描时显示各版本的记录

可以通过指定 VERSIONS=>3 显示最近三次修改的记录，使用 RAW 显示操作信息。在修改数据时，每一次都会记录一个 timestamp，即当时的修改时间。

```
hbase(main):034:0> scan "stud",{RAW=>true,VERSIONS=>3,COLUMNS=>"info"}
ROW     COLUMN+CELL
U001    column=info:age, timestamp=1500783145216, value=23
U001    column=info:name, timestamp=1500783878309, value=ThirdName
U001    column=info:name, timestamp=1500783873566, value=SecondName
U001    column=info:name, timestamp=1500783868549, value=FirstName
...
row(s) in 0.0480 seconds
```

通过上面的查询，可以看到 VERSIONS=>3 显示了最近三次操作的记录。timestamp 为操作时间，以倒序显示。

（4）扫描过滤

也可以在扫描时使用过滤功能。比如指定从哪一个 Row Key 开始，则可以使用 startrow 过滤：

```
hbase(main):036:0> scan "stud",{COLUMNS=>"info",STARTROW=>"U002"}
ROW     COLUMN+CELL
U002      column=info:name, timestamp=1500783317379, value=Mary
row(s) in 0.0240 seconds
```

也可以使用 ENDROW 指定结束的 Row Key，但请注意，ENDROW 的值是不包含的，即如果指定 ENDROW 行键的值为 U003 且包含 U003，则应该指定 ENDROW=>"U004"：

```
hbase(main):043:0>scan stud",{COLUMNS=>"info",STARTROW=>"U002",ENDROW=>"U004"}
ROW     COLUMN+CELL
U002      column=info:name, timestamp=1500783317379, value=Mary
```

```
U003    column=info:name, timestamp=1500784918046, value=Alex
2 row(s) in 0.0340 seconds
```

上面显示的范围为>=U002 且< U004。

（5）值过滤

可以使用 ValueFilter 实现值过滤功能。因为 HBase 中数据是以二进制形式保存的，所以比较方式为 binary 即二进制。比如查询值等于 Mary 的记录：

```
hbase(main):003:0> scan "stud",FILTER=>"ValueFilter(=,'binary:Mary')"
ROW         COLUMN+CELL
U002        column=info:name, timestamp=1500783317379, value=Mary
```

上面显示值为 Mary 的记录，值得说明的是，如果多个列的值为 Mary，则会查询出多列的数据。值比较与列名无关。

（6）包含

包含类似于 contains，如果值中包含某个字符串，则可以查询出相关的记录。比如查询在值里面包含 Mary 的记录：

```
hbase(main):005:0>scan \
"stud",FILTER=>"ValueFilter(=,'substring:Mary')"
ROW     COLUMN+CELL
U002    column=info:name, timestamp=1500783317379, value=Mary
U004    column=info:desc, timestamp=1500790561908, value=Mary
...
```

（7）列名过滤

可以使用 ColumnPrefixFilter 只查询某些指定的列，比如显示所有 name 的列：

```
hbase(main):017:0> scan "stud",FILTER=>"ColumnPrefixFilter('name')"
ROW     COLUMN+CELL
U001    column=info:name, timestamp=1500783878309, value=ThirdName
U002    column=info:name, timestamp=1500783317379, value=Mary
U003    column=info:name, timestamp=1500784918046, value=Alex
```

使用 ColumnPrefixFilter 时，过滤的是列族后面的列名，而不是列族的名称。

（8）多个条件进行组合

使用 AND 或 OR 关键字，可以串联多个过滤条件。比如查询列名为 name，且值中包含 Mary 的记录：

```
hbase>scan stud",FILTER=>"ColumnPrefixFilter('name') AND ValueFilter(=,'substring:Mary')"
ROW     COLUMN+CELL
U002    column=info:name, timestamp=1500783317379, value=Mary
```

（9）行键过滤

可以使用 PrefixFilter 实现 Row Key 的过滤。比如只查询 Row Key 以 U001 开始的记录：

```
hbase(main):020:0> scan "stud",FILTER=>"PrefixFilter('U001')"
ROW     COLUMN+CELL
U001    column=info:age, timestamp=1500783145216, value=23
```

```
 U001    column=info:name, timestamp=1500783878309, value=ThirdName
```

（10）数据查询

语法：get "表名","行键"[,"列族:[列名]]

查询语句中，行键是必须存在的，列族和列名可以省略。

以下查询行键的值为 U001 的记录：

```
hbase(main):022:0> get "stud","U001"
COLUMN      CELL
 info:age    timestamp=1500783145216, value=23
 info:name   timestamp=1500783878309, value=ThirdName
```

以下只查询行键的值为 U001，且列族名称为 info 的记录：

```
hbase(main):023:0> get "stud","U001","info"
COLUMN      CELL
 info:age    timestamp=1500783145216, value=23
 info:name   timestamp=1500783878309, value=ThirdName
```

以下查询行键的值为 U001，且列的名称为 info:name 的记录：

```
hbase(main):024:0> get "stud","U001","info:name"
COLUMN      CELL
 info:name   timestamp=1500783878309, value=ThirdName
```

（11）修改数据

如果存在相同的数据，则 put 为修改数据：

```
hbase(main):020:0> put "stud","rk001","info:name","Jerry"
```

（12）删除数据

语法：deleteall "表名","行键","列族:列名"

```
hbase(main):083:0> delete "stud","rk001","info:age"
```

（13）删除整个行键中的所有数据

语法：deleteall "表名","行键"

```
hbase(main):087:0> deleteall "stud","rk001"
0 row(s) in 0.0190 seconds
```

（14）删除整个表中的所有数据

语法：truncate "表名"

```
hbase(main):090:0> truncate "stud"
```

HBases 还有更多的操作命令，在此就不再一一赘述。读者想深入了解上面的知识，可以通过查看 HBase 的 API 官方文档，获取所有命令的使用方式。

9.5 本章小结

本章主要讲解了以下内容：

- HBase 指 Hadoop DataBase，它是面向列的数据库，具有如下特点：
  - ➢ 大，一个表可以保存上亿级别的数据。
  - ➢ 利用 HDFS 实现分布式的存储。
  - ➢ 保存的数据，都是二进制形式，没有数据类型。
- HBase 的主要进程是 HMaster 和 HRegionServer。
- HBase 伪分布配置时可以使用 HBase 内置的 ZooKeeper，也可以使用外部独立的 ZooKeeper，只要在 hbase-env.sh 中修改 HBASE_MANAGER_ZK=false 且在 hbase-site.xml 中添加配置 hbase.zookeeper.quorum=ip:2181 即可。
- Java 代码连接 HBase，无论是伪分布式或是真分布式，只要配置连接到 ZooKeeper，比如：hbaseConfiguration.set("hbase.zookeeper.quorum","ip:2181");即可成功连接 HBase。但在连接之前一定要配置 hosts 文件，指定连接的 IP 地址中的主机名与 IP 地址的对应关系。

第 10 章

Flume 数据采集

本章主要内容：

- Flume 简介。
- Flume 的安装与配置。
- Flume 部署。

Flume 是 Cloudera 公司提供的一个高可用、高可靠、分布式的海量日志采集、聚合和传输的系统。Flume 支持定制各类数据源如 Avro、Thrift、Spooling 等。同时 Flume 提供对数据的简单处理，并将数据处理结果写入各种数据接收方，比如将数据写入到 HDFS 文件系统中。

10.1 Flume 简介

Flume 作为 Cloudera 公司开发的实时日志收集系统，受到了业界的认可与广泛应用。2010 年 11 月 Cloudera 开源了 Flume 的第一个可用版本 0.9.2，这个系列版本被统称为 Flume-OG（Original Generation）。随着 Flume 功能的扩展，Flume-OG 代码开始臃肿、核心组件设计不合理、核心配置不标准等缺点暴露出来，尤其是在 Flume-OG 的最后一个发行版本 0.94.0 中，日志传输不稳定的现象尤为严重。为了解决这些问题，2011 年 10 月，Cloudera 重构了核心组件、核心配置和代码架构，重构后的版本统称为 Flume-NG（Next Generation）。重构的另一原因是将 Flume 纳入 Apache 旗下，Cloudera Flume 改名为 Apache Flume。

10.1.1 Flume 原理

Flume 的数据流由事件（Event）贯穿始终。事件是 Flume 的基本数据单位，它携带日志数据（字节数组形式）并且携带有头信息，这些 Event 由 Agent 外部的 Source 生成，当 Source 捕获事件后会进行特定的格式化，然后 Source 会把事件推入（单个或多个）Channel 中。可以把 Channel 看作是一个缓冲区，它将保存事件直到 Sink 处理完该事件。Sink 负责持久化日志或者把事件推向另一个 Source。

Flume 以 Agent 为最小的独立运行单位。一个 Agent 就是一个 JVM。Agent 由 Source、Sink 和 Channel 三大组件构成，如图 10-1 所示。

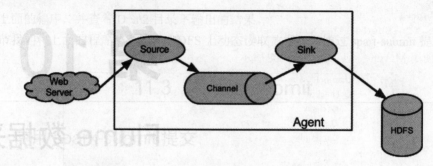

图 10-1

值得注意的是，Flume 提供了大量内置的 Source、Channel 和 Sink。不同类型的 Source、Channel 和 Sink 可以自由组合。组合方式基于用户的配置文件，非常灵活。比如，Channel 可以把事件暂存在内存里，也可以持久化到本地硬盘上。如图 10-2 所示。Sink 可以把日志写入 HDFS、HBase，甚至是另外一个 Source 等。Flume 支持用户建立多级流，也就是说多个 Agent 可以协同工作，并且支持 Fan-in（扇入）、Fan-out（扇出）、Contextual Routing、Backup Routes。

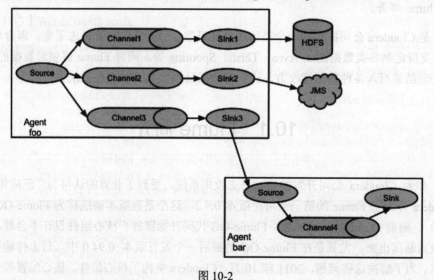

图 10-2

10.1.2　Flume 的一些核心概念

Agent 使用 JVM 运行 Flume。每台机器运行一个 Agent，但是可以在一个 Agent 中包含多个 Sources 和 Sinks。

Source 从 Client 收集数据，传递给 Channel。

Channel 连接 Sources 和 Sinks，Channel 缓存从 Source 收集来的数据。

Sink 从 Channel 收集数据,并将数据写到目标文件系统中。

10.2 Flume 的安装与配置

在安装 Flume 之前,需要确认已经安装了 JDK 并正确配置了环境变量。

步骤 01 下载并解压 Flume。下载地址为 http://www.apache.org/dyn/closer.lua/flume/1.9.0/apache-flume-1.9.0-bin.tar.gz。

解压:

```
$ tar -zxvf ~/apache-flume-1.9.0-bin.tar.gz -C /app/
```

步骤 02 配置 flume-env.sh 文件。

在 flume-env.sh 文件中配置 JAVA_HOME 环境变量:

```
$ cp flume-env.sh.template flume-env.sh
$ vim flume-env.sh
export JAVA_HOME=/usr/local/java/jdk1.8.0_211
```

步骤 03 配置 Flume 的环境变量。

```
export FLUME_HOME=/home/isoft/app/flume-1.9.0
export PATH=$PATH:$FLUME_HOME/bin
```

使环境变量生效:

```
$ source ~/.bash_profile
```

现在可以使用 version 测试 Flume 的版本:

```
$ flume-ng version
flume 1.9.0
```

至此,Flume 安装与配置已经完成,非常简单。接下来,我们将部署两个基本的 Flume Agent 来测试 Flume。

10.3 快速示例

根据官网示例,我们做一个快速的 Flume 示例,这个示例从网络上收集数据,并输出到日志中。

步骤 01 定义配置文件。

在任意的目录下定义一个资源文件,如 agent_1.conf,并写入以下内容,其中#开始的语句行为注释,可以不用输入。

```
# 定义三个核心组件
a1.sources = r1
```

```
a1.channels = c1
a1.sinks = k1
# 定义 sources 即头 1 的类型及绑定的端口
a1.sources.r1.type = netcat
a1.sources.r1.bind = localhost
a1.sources.r1.port = 44444
# 定义输出的目标为系统日志
a1.sinks.k1.type = logger
# 定义 channel 的类型及大小
a1.channels.c1.type = memory
a1.channels.c1.capacity = 1000
a1.channels.c1.transactionCapacity = 100
# 将三个对象的关系进行整合
a1.sources.r1.channels = c1
# 注意以下 channel 后面没有 s
a1.sinks.k1.channel = c1
```

步骤02 启动 Agent。

使用 flume-ng 命令，即可启动一个 Agent，通过 --conf-files 可以指定自己前面配置的配置文件。

```
$ /app/flume-1.9.0/bin/flume-ng agent --conf /home/isoft/app/flume-1.9.0/conf
 --conf-file /app/conf/flume/agent_1.conf \
 --name a1 -Dflume.root.logger=INFO,console
```

说明：

- --conf：用于指定 Flume 配置文件的目录，这个目录下包含 flume-env.sh 等文件。
- --conf-file：用于指定用户自己配置的 Agent 文件。
- --name：用于指定用户配置文件中的 Agent 名称。
- -Dxx：用于配置系统一些变量（如日志）。注意，配置中 console 的 c 是小写的。

步骤03 访问 44444 端口并输入数据。

```
[hadoop@server201 ~]$ telnet localhost 44444
Trying ::1...
telnet: connect to address ::1: Connection refused
Trying 127.0.0.1...
Connected to localhost.
Escape character is '^]'.
Jack
Mary
Rose
```

查看 flume-ng 的控制台日志：

```
{ headers : {} body : A4 61 63 6B 0D       Jack. }
{ headers : {} body : A4 61 63 6B 0D       Mary. }
{ headers : {} body : A4 61 63 6B 0D       Rose. }
```

至此，一个简单的 Flume agent 就算是配置完成了。

10.4 在 ZooKeeper 中保存 Flume 的配置文件

我们也可以将 Flume 配置文件存入 ZooKeeper 中，首先看一下 Flume 在 ZooKeeper 节点中的结构，根据官方文档的说明，Flume 在 ZooKeeper 节点的结构类似以下结果：

```
- /flume
 |- /a1 [Agent config file]
 |- /a2 [Agent config file]
```

首先启动 Agent，使用-z 指定 ZooKeeper 的主机地址，使用-p 指定 ZooKeeper 的节点名称。通过指定-z、-p 两个参数，使用 ZooKeeper 中的节点信息。启动命令如下：

```
$ bin/flume-ng agent -conf conf -z zkhost:2181,zkhost1:2181 -p /flume -name a1 -Dflume.root.logger=INFO,console
```

这两个参数的具体含义如下：

- z：默认为-，ZooKeeper 连接字符串。逗号分隔的主机名列表：端口。
- p：默认/flume，ZooKeeper 中用于存储代理配置的基本路径。

为了可以将 Flume 的配置文件写入到 ZooKeeper 节点的 data 中，我们需要编写一个 Java 类，然后在 Linux 上使用 Java 命令，将指定的文件上传到 zk 的节点数据中去。

步骤 01 开发 ZooKeeper 上传文件的程序。

添加依赖：

```xml
<dependency>
    <groupId>org.apache.zookeeper</groupId>
    <artifactId>zookeeper</artifactId>
    <version>3.6.2</version>
</dependency>
```

【代码 10-1】ZkUpload.java

```
1.  package org.hadoop.flume;
2.  public class ZkUpload {
3.      /**
4.       * 上传文件
5.       */
6.      public static void main(String[] args) throws Exception {
7.          if (args.length < 3) {
8.              System.err.println("用法: <zkServer> <path> <file>");
9.              return;
10.         }
11.         String zkHosts = args[0];
12.         String path = args[1];
13.         String file = args[2];
14.         CountDownLatch countDownLatch = new CountDownLatch(1);
15.         ZooKeeper zooKeeper = new ZooKeeper(zkHosts, 3000, new Watcher() {
16.             @Override
```

```
17.        public void process(WatchedEvent event) {
18.            System.out.println("连接" + zkHosts + "成功");
19.            countDownLatch.countDown();
20.        }
21.    });
22.    countDownLatch.await();
23.    try {
24.        //检查文件是否存在
25.        Stat stat = zooKeeper.exists(path, null);
26.        if (stat != null) {
27.            throw new RuntimeException("节点已经存在！");
28.        }
29.        //开始上传文件
30.        InputStream in = new FileInputStream(file);
31.        //判断文件大小，最大是1MB
32.        int size = in.available();
33.        if (size > (1024 * 1024)) {
34.            throw new RuntimeException("上传的文件大小不能超过1M");
35.        }
36.        ByteArrayOutputStream bout = new ByteArrayOutputStream();
37.        byte[] bs = new byte[1024 * 4];
38.        int len = 0;
39.        while ((len = in.read(bs)) != -1) {
40.            bout.write(bs, 0, len);
41.        }
42.        bout.close();
43.        bs = bout.toByteArray();
44.        path = zooKeeper.create(path, bs, ZooDefs.Ids.OPEN_ACL_UNSAFE, CreateMode.PERSISTENT);
45.        System.out.println("节点创建成功：" + path);
46.    } finally {
47.        zooKeeper.close();
48.    }
49. }
50. }
```

步骤 02 现在将代码上传到 Linux 服务器并编译。

上传代码，并直接使用 javac 编译 ZkUpload.java：

```
$ javac -classpath ./app/zookeeper-3.6.2/lib/* -d . ZkUpload.java
```

运行测试：

```
[hadoop@server201 java]$ java org.hadoop.flume.ZkUpload
   用法： <zkServer> <path> <file>
```

如果显示了用法，则说明代码可用。

步骤 03 启动 ZooKeeper 并上传配置文件。

```
$ java -jar zk.jar org.hadoop.flume.ZkUpload localhost:2181 /flume/a1 /app/datas/conf/flume/agent_1.conf
```

查看 ZooKeeper 上的/flume/a1 里面的内容：

```
[zk: localhost:2181(CONNECTED) 3] ls /flume
[a1]
```

可以获得以下内容，可见配置文件已经上传到节点的数据中：

```
[zk: localhost:2181(CONNECTED) 4] get /flume/a1
# 定义三个核心组件
a1.sources = r1
a1.channels = c1
a1.sinks = k1
# 定义 sources 即头 1 的类型及绑定的端口
a1.sources.r1.type = netcat
a1.sources.r1.bind = localhost
a1.sources.r1.port = 44444
# 定义输出的目标为系统日志
a1.sinks.k1.type = logger
# 定义 channel 的类型及大小
a1.channels.c1.type = memory
a1.channels.c1.capacity = 1000
a1.channels.c1.transactionCapacity = 100
# 将三个对象的关系进行整合
a1.sources.r1.channels = c1
# 注意以下 channel 后面没有 s
a1.sinks.k1.channel = c1
```

步骤 04 为 Flume 添加 ZooKeeper 依赖。

由于启动 flume-agent 时需要读取 ZooKeeper 中的数据，此时需要依赖 ZooKeeper。根据 Flume 官方提示，可以在$FLUME_HOME 目录下创建 plugins.d 目录，并将依赖保存到这个目录下。plugins.d 的目录结构应该为：

```
plugins.d/
plugins.d/custom-source-1/
plugins.d/custom-source-1/lib/my-source.jar
plugins.d/custom-source-1/libext/spring-core-2.5.6.jar
plugins.d/custom-source-2/
plugins.d/custom-source-2/lib/custom.jar
plugins.d/custom-source-2/native/gettext.so
```

目录结构说明如下：

- lib：依赖的主要 jar 文件。
- libext：依赖的扩展 jar 文件。
- native：本地库如 c 语言编译的.so 文件。

现在我们将 ZooKeeper 的所有 jar 包放到 flume_home/plugins.d/zookeeper/lib 目录下：

```
$ ln -s /home/isoft/app/zookeeper-3.5.5/lib/* \
 /home/isoft/app/flume-1.9.0/plugins.d/zookeeper/lib
```

步骤 05 启动 flume-agent。

```
$ flume-ng agent --conf /app/app/flume-1.9.0/conf/  \
-z server201:2181 \
-p /flume -name a1  \
-Dflume.root.logger=INFO,console
```

启动成功，将显示以下信息：

```
Starting Channel c1
Starting Sink k1
Starting Source r1
Created serverSocket:sun.nio.ch.ServerSocketChannelImpl[/127.0.0.1:44444
```

至此，就可以使用 ZooKeeper 集群中配置的文件了。

10.5　Flume 的更多 Source

所有 Flume 的 Source 都可以在官网上找到它的配置。

10.5.1　avro source

avro source 可以定制 avro-client 发送一个指定的文件给 Flume Agent，avro source 使用 Avro RPC 机制，Flume 主要的 RPC Source 也是 avro source，它使用 Netty-Avro inter-process 的通信（IPC）协议来通信，因此可以用 Java 或 JVM 语言发送数据到 avro source 端。它的配置文件主要包含三个参数：

- type：avro source 的别名是 avro，也可以使用完整的类别名称 org.apache.flume.source. AvroSource。
- bind：绑定的 IP 地址或主机名。若不指定端口号，则表示绑定机器的所有端口。
- port：绑定监听端口。

1. 通过 avro-client 发送文件数据

flume-ng 自带 avro-client 命令，可以将指定的文件通过 avro rpc 发送到指定的 avro 源。
当输入 flume-ng help 命令后，就可以看到这个 avro-client 命令：

```
[hadoop@server201 flume-1.9.0]$ bin/flume-ng help
Usage: bin/flume-ng <command> [options]...
commands:
  help                      display this help text
  agent                     run a Flume agent
  avro-client               run an avro Flume client
  version                   show Flume version info
```

avro source 配置示例：

```
#配置三个源
```

```
a1.sources=r1
a1.channels=c1
a1.sinks=k1
#配置 r1 的输入源为 avro
a1.sources.r1.type=avro
a1.sources.r1.bind=0.0.0.0
a1.sources.r1.port=4141
#配置 channels 的类型为 memory
a1.channels.c1.type=memory
a1.channels.c1.capacity=1000
a1.channels.c1.transactionCapacity=100
#配置 sinks 的目标为 logger
a1.sinks.k1.type=logger
#组织三个组件
a1.sources.r1.channels=c1
a1.sinks.k1.channel=c1
```

现在启动 avro source：

```
$ flume-ng agent -n a1 -c cont/ --conf-file conf/01_avro.conf \
-Dflume.root.logger=INFO,consol
```

然后使用 flume-ng avro-client 向 4141 端口发送文件数据：

```
$flume-ng avro-client -c conf/ -H 127.0.0.1 -p 4141 -F hello.txt
```

现在就可以查看日志 log/flume.log 中的内容：

```
{ headers : {} body : ...}
{ headers : {} body : ...}
{ headers : {} body : ...}
```

2. Java api 向 avroi 源发送数据

同样地，也可以使用 Java 代码向 avro 源发送数据。可以通过官方地址获取参考代码。官方地址为 http://flume.apache.org/releases/content/1.9.0/FlumeDeveloperGuide.html#rpc-clients-avro-and-thrift。

使用上面的配置文件，并启动 agent。

首先，需要添加依赖：

```
<dependency>
    <groupId>org.apache.flume</groupId>
    <artifactId>flume-ng-core</artifactId>
    <version>1.9.0</version>
</dependency>
```

开发一个基本的测试，此测试向 Flume 监听的端口发送数据。代码如下所示。

【代码 10-2】SendData.java

```
1. package org.hadoop.flume;
2. public class SendData {
3.     @Test
4.     public void sendData() throws Exception {
```

```
5.        String host = "server201";
6.        int port = 4141;
7.        RpcClient client = RpcClientFactory.getDefaultInstance(host, port);
8.        //发送数据
9.        Event event = EventBuilder.withBody("Hello", Charset.forName("UTF-8"));
10.       client.append(event);
11.       client.close();//关闭
12.    }
13. }
```

上面的代码中,通过 client.append 即可把 Event 发送到 avro source 上去。

连续发送 event 的示例如【代码 10-3】所示。

【代码 10-3】SendData2.java

```
1. package org.hadoop.flume;
2. public class SendData2 {
3.    public static void main(String[] args) throws Exception {
4.        String host = "server201";
5.        int port = 4141;
6.        RpcClient client = RpcClientFactory.getDefaultInstance(host, port);
7.        //连续输入数据
8.        Scanner sc = new Scanner(System.in);
9.        while (true) {
10.           String str = sc.nextLine();
11.           if (str.equals("exit")) {
12.              break;
13.           }
14.           Event event = EventBuilder.withBody(str, Charset.forName("UTF-8"));
15.           client.append(event);
16.        }
17.        client.close();
18.    }
19. }
```

启动 Java 程序并连续输入数据,然后查看 Flume 收到的数据如下:

```
{ header : {} bydy : A1 B4 64 A3 Jack}
{ header : {} bydy : A1 B4 64 A3 Jack}
...
```

源代码说明:

在 Flume 1.4 以后,连接 avro source 或是连接 thirft source 都可以使用 RpcClient。RpcClient 的两个主要子类为:ThriftRpcClient 和 nettyAvroRpcClient。如图 10-3 所示。

第 10 章　Flume 数据采集 | 177

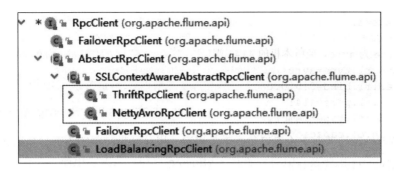

图 10-3

上面的代码，由于连接是 Avro source，所以返回的对象为 NettyAvroRpcClient，即代码 RpcClient client = RpcClientFactory.getDefaultInstance(host,port)返回的是 NettyAvroRpcClient。

3. 开发一个 netcat 源向 avro 发送数据

下面开发一个配置示例，将配置两个 Agent，两个 Agent 之间通过 avro rpc 进行通信。配置如图 10-4 所示。

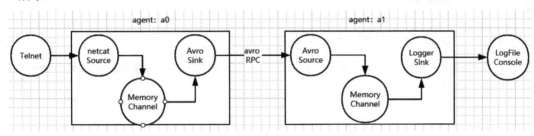

图 10-4

修改 avro.conf 内容如下：

```
#配置 agent a0
a0.sources=r0
a0.channels=c0
a0.sinks=k0
a0.sources.r0.type=netcat
a0.sources.r0.bind=0.0.0.0
a0.sources.r0.port=4040
a0.channels.c0.type=memory
a0.channels.c0.capacity=1000
a0.channels.c0.transactionCapacity=100
#配置 sink 为 avro 并向一个固定的 ip 和端口号 4141
a0.sinks.k0.type=avro
a0.sinks.k0.hostname=127.0.0.1
a0.sinks.k0.port=4141
a0.sources.r0.channels=c0
a0.sinks.k0.channel=c0
############################
#配置 agent a1，内容同上直接 copy
a1.sources=r1
```

```
a1.channels=c1
a1.sinks=k1
#配置source为avro，监听本机的4141端口
a1.sources.r1.type=avro
a1.sources.r1.bind=0.0.0.0
a1.sources.r1.port=4141
a1.channels.c1.type=memory
a1.channels.c1.capacity=1000
a1.channels.c1.transactionCapacity=100
a1.sinks.k1.type=logger
a1.sources.r1.channels=c1
a1.sinks.k1.channel=c1
```

注意，上面的4141就是avro接收数据的端口。

现在启动agent a1：

```
$ bin/flume-ng agent  -n  a1 -c conf/   --conf-file conf/avro.conf \
-Dflume.root.logger=INFO,console
```

然后再启动agent a0：

```
$bin/flume-ng agent  -n   a0 -c conf/   --conf-file    conf/avro.conf \
-Dflume.root.logger=INFO,console
```

现在就可以使用telnet登录，向这个netcat发送数据了，然后这个数据会通过avro rpc发送到avro源，最后输出到日志文件中。

以下是向agent a0发送数据：

```
[hadoop@server201 ~]$ telnet localhost 4040
Trying ::1...
Connected to localhost.
Escape character is '^]'.
Jack
OK
```

以下是agent a1输出到日志的数据：

```
{ headers:{} body: 4A 61 63 6B 0D                                  Jack. }
{ headers:{} body: 4A 61 63 6B 0D                                  Jack. }
{ headers:{} body: 4A 61 63 6B 0D                                  Jack. }
```

至此，完成了两个Agent通过avro rpc通信的演示。

10.5.2　thrift source 和 thrift sink

thrift与avro类似，同样可以进行串联，也都可以实现rpc调用，但thrift支持更多的语言。

1. 使用Java代码向thrift发送数据

与avro类似，thrift也是通过RpcClient向thrift source发送数据的。

首先定义一个Agent配置文件thrift.conf，内容如下：

```
#定义三个组件：
```

```
a1.sources=r1
a1.channels=c1
a1.sinks=k1
#定义thrift source,绑定本机所有ip的4141端口
a1.sources.r1.type=thrift
a1.sources.r1.bind=0.0.0.0
a1.sources.r1.port=4141
#channel依然使用内存
a1.channels.c1.type=memory
a1.channels.c1.capacity=1000
a1.channels.c1.transactionCapacity=100
#sink为日志
a1.sinks.k1.type=logger
#绑定三个组件的关系
a1.sources.r1.channels=c1
a1.sinks.k1.channel=c1
```

现在启动 agent a1:

```
$bin/flume-ng  agent  -n a1 -c conf/   --conf-file conf/thritt.conf    \
-Dflume.root.logger=INFO,console
```

编写 Java 客户端,实现发送数据的功能。通过 RpcClientFactory.getThriftInstance(host,port)方法,可以获取一个 Thrift 的 RpcClient 对象,如【代码 10-4】所示。

【代码 10-4】ThriftClient.java

```
1.  package org.hadoop.flume;
2.  public class ThriftClient {
3.      @Test
4.      public void test1() throws Exception {
5.          String host = "server101";
6.          int port = 4141;
7.          RpcClient client = RpcClientFactory.getThriftInstance(host, port);
8.          System.err.println(client);
9.          Event event = EventBuilder.withBody("Hello", Charset.forName("UTF-8"));
10.          client.append(event);
11.          client.close();//关闭
12.     }
13. }
```

接下来开发一个连续发送数据的测试,如【代码 10-5】所示。

【代码 10-5】ThriftSender.java

```
1.  package org.hadoop.flume;
2.  public class ThriftSender {
3.      public static void main(String[] args) throws Exception {
4.          String host = "server201";
5.          int port = 4141;
6.          RpcClient client = RpcClientFactory.getThriftInstance(host,port);
7.          //连续输入数据
8.          Scanner sc = new Scanner(System.in);
```

```
9.        while(true){
10.           String str = sc.nextLine();
11.           if(str.equals("exit")){
12.              break;
13.           }
14.           Event event = EventBuilder.withBody(str, Charset.forName("UTF-8"));
15.           client.append(event);
16.        }
17.        client.close();
18.     }
19. }
```

运行【代码10-5】,并传入一些数据,查看 Flume 日志接收到的数据:

```
{ headers:{} body: 48 65 6C 6C 6F                              Hello }
{ headers:{} body: 48 65 6C 6C 6F                              First }
{ headers:{} body: 48 65 6C 6C 6F                              Second }
...
```

2. thrift sink

类似于 avro,也可以将多个 thrift 进行串联,即 thrift sink 的输出为 thrift source 的输入。配置如图10-5所示。

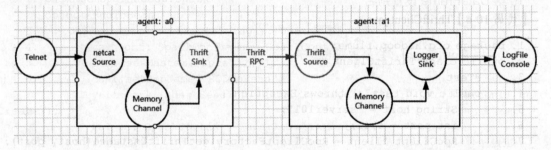

图10-5

配置文件 thrift.conf:

```
#配置 agent a0:
a0.sources=r0
a0.channels=c0
a0.sinks=k0
a0.sources.r0.type=netcat
a0.sources.r0.bind=0.0.0.0
a0.sources.r0.port=4040
a0.channels.c0.type=memory
a0.channels.c0.capacity=1000
a0.channels.c0.transactionCapacity=100
#定义 thrift 的输出
a0.sinks.k0.type=thrift
a0.sinks.k0.hostname=127.0.0.1
a0.sinks.k0.port=4141
```

```
#定义 agent a0 关联的三个组件
a0.sources.r0.channels=c0
a0.sinks.k0.channel=c0
############定义 agent a1,即第二个 agent
a1.sources=r1
a1.channels=c1
a1.sinks=k1
#定义 thrift 的 souce 即输入,注意这个端口号与上面 thrift sink 的端口号相同
a1.sources.r1.type=thrift
a1.sources.r1.bind=0.0.0.0
a1.sources.r1.port=4141
a1.channels.c1.type=memory
a1.channels.c1.capacity=1000
a1.channels.c1.transactionCapacity=100
a1.sinks.k1.type=logger
#关联 agent a1 的三个组件
a1.sources.r1.channels=c1
a1.sinks.k1.channel=c1
```

启动 agent a1:

```
$ flume-ng agent  -n a1 -c conf/  --conf-file conf/thrift.conf \
-Dflume.root.logger=INFO,console
```

启动 agent a0:

```
$ flume-ng agent  -n a0 -c conf/  --conf-file conf/thrift.conf  \
-Dflume.root.logger=INFO,console
```

现在通过 telnet 访问 agent a0 的 4040 端口:

```
$ telnet       127.0.0.1   4040
Jack
ok
Mary
ok
```

查看 flume agent a1 输出的日志信息:

```
{ headers:{} body: 4A 61 63 6B 0D                      Jack.}
{ headers:{} body: 4A 61 63 6B 0D                      Mary.}
```

10.5.3 exec source

exec source 可以执行 Unix 命令,比如 tail -f somefile.log。
定义 exec.conf 配置文件,内容如下:

```
#定义三个组件
a1.sources=r1
a1.channels=c1
a1.sinks=k1
#定义 exec source,执行的命令通过 command 定义
a1.sources.r1.type=exec
```

```
a1.sources.r1.command=tail -F /home/isoft/a.log
a1.channels.c1.type=memory
a1.channels.c1.capacity=1000
a1.channels.c1.transactionCapacity=100
a1.sinks.k1.type=logger
#组合三个组件
a1.sources.r1.channels=c1
a1.sinks.k1.channel=c1
```

启动 flume agent：

```
$ bin/flume-ng agent     -n a1 -c conf/  --conf-file conf/exec.conf \
-Dflume.root.logger=INFO,console
```

向 a.log 日志文件中追加数据：

```
$echo a >> a.log
$echo Jack >> a.log
$echo Mary >> a.log
```

查看日志的输出：

```
{ headers:{} body: 61                                                    a }
{ headers:{} body: 61                                                 Jack }
{ headers:{} body: 61                                                 Mary }
```

10.5.4　spool source

spool source 用于实时监控一个目录下新增的文件，主要是文本文件。

创建 flume agent 配置文件 spool.conf，内容如下：

```
#声明三个组件
a1.sources=r1
a1.channels=c1
a1.sinks=k1
#定义 spool source，读取指定的目录
a1.sources.r1.type=spooldir
a1.sources.r1.spoolDir=/home/isoft/a
a1.sources.r1.fileHeader=true
a1.channels.c1.type=memory
a1.channels.c1.capacity=1000
a1.channels.c1.transactionCapacity=100
a1.sinks.k1.type=logger
#组织这三个组件
a1.sources.r1.channels=c1
a1.sinks.k1.channel=c1
```

启动这个 agent：

```
$ flume-ng agent   -n a1 -c conf/   --conf-file conf/spool.conf \
-Dflume.root.logger=INFO,console
```

将任意的一个文本文件直接 copy 到/home/isoft/a 目录下：

```
$ cp a.txt /home/hadoop/a/
```

然后，现在看 flume agent 输出的日志，由于设置了 fileHeader=true，所以会携带 header 信息：

```
{ headers:{file=/home/isoft/a/a.log} body: 61                a }
{ headers:{file=/home/isoft/a/a.log} body: 61 62 63          abc }
{ headers:{file=/home/isoft/a/a.log} body: 61 62 63          Jack }
{ headers:{file=/home/isoft/a/a.log} body: 61 62 63          Mike }
{ headers:{file=/home/isoft/a/a.log} body: 61 62 63          Rose }
```

10.5.5 HDFS sinks

定义配置文件 hdfs.conf，内容如下：

```
#定义三个组件
a1.sources=r1
a1.channels=c1
a1.sinks=k1
a1.sources.r1.type=netcat
a1.sources.r1.bind=0.0.0.0
a1.sources.r1.port=4141

a1.channels.c1.type=memory
a1.channels.c1.capacity=1000
a1.channels.c1.transactionCapacity=100
#定义 hdfs sink
a1.sinks.k1.type=hdfs
#输出的 hdfs 目录，/flume/events 目录必须已经存在
a1.sinks.k1.hdfs.path=/flume/events/%Y%m%d/%H%M/%S
a1.sinks.k1.hdfs.filePrefix=FlumeData
#每 10 分钟一个子目录
a1.sinks.k1.hdfs.round=true
a1.sinks.k1.hdfs.roundUnit=minute
a1.sinks.k1.hdfs.roundValue=10
a1.sinks.k1.hdfs.useLocalTimeStamp=true   #必须设置为 true 才可使用%Y
a1.sinks.k1.hdfs.writeFormat=Text   #以文本形式输出，否则为字节码
#组织三个组件
a1.sources.r1.channels=c1
a1.sinks.k1.channel=c1
```

启动这个 agent：

```
$ flume-ng       agent     -n  a1  -c  conf/   --conf-file conf/hdfs.conf \
-Dflume.root.logger=INFO,console
```

然后通过 telnet 登录：

```
[hadoop@server201 ~]$ telnet 127.0.0.1 4141
Trying 127.0.0.1...
Connected to 127.0.0.1.
Escape character is '^]'.
Smith
Rose
```

通过-text 查看内容：

```
[hadoop@server201 ~]$ hdfs dfs -text /flume/events/2020310/0440/00/*
1570696963767   Smith
1570697055723   Rose
```

可见，内容已经添加到 HDFS 文件中去了。

10.6　本章小结

本章主要讲解了以下内容：

- Flume 是数据采集工具，主要用于采集日志信息。
- Agent 是 Flume 的核心，每一个 Agent 都包含 Source、Channel 和 Sink。
- 多个 Agent 可以组成一个链。
- Flume 定义了丰富的组件，只需要在配置文件中配置并组合它们即可。

第 11 章

Spark 框架搭建及应用

本章主要内容：

- Spark 的安装与运行。
- Spark On YARN 运行模式。
- 使用 Scala 开发 Spark 应用。
- Spring 算子。
- Spark SQL。
- Spark Streaming。

Spark 的核心是建立在统一的抽象弹性分布式数据集（Resiliennt Distributed Datasets，RDD）之上的，这使得 Spark 的各个组件可以无缝地进行集成，能够在同一个应用程序中完成大数据处理。RDD 是 Spark 提供的最重要的抽象概念，它是一种有容错机制的特殊数据集合，可以分布在集群的结点上，以函数式操作集合的方式进行各种并行操作。

通俗来讲，可以将 RDD 理解为一个分布式对象集合，本质上是一个只读的分区记录集合。每个 RDD 可以分成多个分区，每个分区就是一个数据集片段。一个 RDD 的不同分区可以保存到集群中的不同结点上，从而可以在集群中的不同结点上进行并行计算。

RDD 具有容错机制，并且只读不能修改，可以执行确定的转换操作创建新的 RDD。具体来讲，RDD 具有以下几个属性。

- 只读：不能修改，只能通过转换操作生成新的 RDD。
- 分布式：可以分布在多台机器上进行并行处理。
- 弹性：计算过程中内存不够时它会和磁盘进行数据交换。
- 基于内存：可以全部或部分缓存在内存中，在多次计算间重用。

RDD 的操作分为转化（Transformation）操作和行动（Action）操作，转化操作又叫作转换算子。转化操作就是从一个 RDD 产生一个新的 RDD，而行动操作就是进行实际的计算。RDD 的操作是惰性的，当 RDD 执行转化操作的时候，实际计算并没有被执行，只有当 RDD 执行行动操作时，才会促发计算任务提交,从而执行相应的计算操作。

Spark 与 Hadoop 紧密集成,它可以在 YARN 上运行,并支持 Hadoop 文件格式。Spark 最突出的表现在于它将数据集存储在内存中。Spark 的 DAG 引擎与 MapReduce 不同,它可以处理任意操作流水线,并为用户转换为单个作业。Spark 提供了三种语言的 API,分别是 Java、Scala 和 Python。本章所使用的语言大多为 Scala。不过,在 Spark 的官方网站上,可以同时找到 Java 和 Scala 两个版本的代码。对于 Scala 这门语言,读者可以通过其他途径学习。

首先下载 Spark。在 Spark 的官方网站 http://spark.apache.org/下载 Spark 二进制包,选择 3.1 版本及对应的 Hadoop 3.2 版本,如图 11-1 所示。

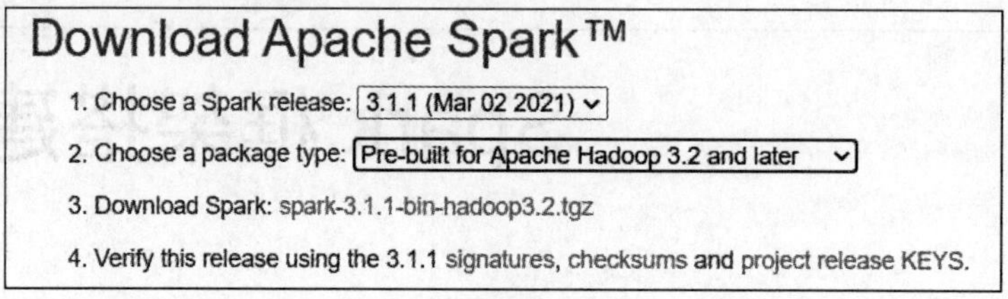

图 11-1

11.1 安装 Spark

Spark 的运行模式不一定必须拥有 Hadoop 环境。Spark 的运行模式可以分为:

- 本地模式。
- 伪分布式。
- 完全分布式及运行在 YARN 上。

无论哪一种运行模式都需要 JDK1.8+环境。所以,需要事先准备好 JDK 环境,并正确配置 JAVA_HOME 和 PATH 环境变量。

11.1.1 本地模式

这种模式比较简单,解压安装包就可以运行。这种运行模式可以让我们快速了解 Spark,我们解压安装包并配置环境变量如下:

```
$ tar -zxvf spark-3.1.1-bin-hadoop3.2.tgz -C /app/
$ sudo vim /etc/profile
export SPARK_HOME=/app/spark-3.1.1
export PATH=$PATH:$SPARK_HOME/bin
$ source /etc/profile
```

配置完成以后,先通过 spark-shell 查看帮助和版本信息。可以通过--help 参数,查看所有帮助:

```
[hadoop@server201 app]$ spark-shell --help
Usage: ./bin/spark-shell [options]
Scala REPL options:
```

```
        -I <file>               preload <file>, enforcing line-by-line interpretation
Options:
  --master MASTER_URL      spark://host:port, mesos://host:port, yarn,
                           k8s://https://host:port, or local (Default: local[*]).
...
```

查看 Spark 的版本直接通过 --vesion 参数即可。其中显示 Spark 的版本为 3.1.1，Scala 的版本为 2.12。

```
[hadoop@server201 app]$ spark-shell --version
Spark Version 3.1.1
Using Scala version 2.12.10, Java HotSpot(TM) 64-Bit Server VM, 1.8.0_281
Branch HEAD
Compiled by user ubuntu on 2021-02-22T01:33:19Z
Revision 1d550c4e90275ab418b9161925049239227f3dc9
Url https://github.com/apache/spark
Type --help for more information.
```

使用 spark-shell 启动 Spark 客户端，以 local 模式运行。通过 --master 指定为 local 模式，通过 local[2] 指定使用两核。

```
$ spark-shell --master local[2]
Welcome to
      ____              __
     / __/__  ___ _____/ /__
    _\ \/ _ \/ _ `/ __/  '_/
   /___/ .__/\_,_/_/ /_/\_\   version 3.1.1
      /_/

Using Scala version 2.12.10 (Java HotSpot(TM) 64-Bit Server VM, Java 1.8.0_28
1)
Type in expressions to have them evaluated.
Type :help for more information.
scala>
```

以下我们运行一个 WordCount 的示例，可能存在一些方法大家尚不明白，不过没有关系，后面的章节中我们将会详细讲解。

首先通过 sc 变量，即 SparkContext 对象，加载一个文件到内存中：

```
scala> val file = sc.textFile("file:///home/hadoop/notes.txt");
file: org.apache.spark.rdd.RDD[String] = file:///home/hadoop/notes.txt MapPartitionsRDD[1] at textFile at <console>:24
```

然后使用一系列的算子对文件对象进行处理，先按空格和回车键进行 split，然后再使用 map 将数据组合成（key,value）形式，最后使用 reduceByKey 算子，将 key 进行合并：

```
scala> val words = file.flatMap(_.split("\\s+")).map((_,1)).reduceByKey(_+_);
words: org.apache.spark.rdd.RDD[(String, Int)] = ShuffledRDD[4] at reduceByKey at <console>:25
```

最后使用 collect 输出结果：

```
scala> words.collect
res0: Array[(String, Int)] = Array((this,2), (is,1), (how,1), (into,2), (some
thing,1), (hive.,2), (file,1), (And,1), (process,1), (you,2), (about,1), (wordco
unt,1), (import,1), (a,1), (text,1), (be,1), (to,2), (in,1), (tell,1), (for,1),
(must,1))
```

上面的运算过程，也可以通过访问 http://server201:4040 查看 DAG 运行效果，如图 11-2 所示。

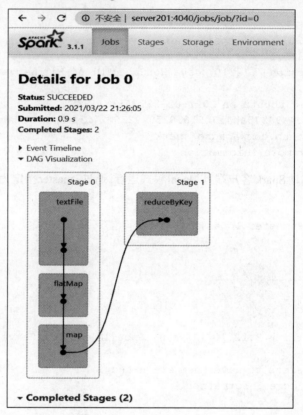

图 11-2

从图 11-2 可以看出，reduceByKey 引发了第二个 Stage，从 Stage0 到 Stage1 将会引发 shuffle。这也是区分转换算子和行动算子的主要依据。

通过上面的示例可以看出，独立运行模式下不需要事先启动任何的进程。启动 spark-shell 后，可以通过 SparkContext 读取本地系统目录下的文件。

操作完成以后，输入 :quit 即可以退出。

```
scala> :quit
[hadoop@server201 app]$
```

11.1.2 伪分布式安装

伪分布式即在一台主机上运行，但需要启动 Spark 的两个进程：Master 和 Worker。启动后，可通过 8080 端口，查看 Spark 的运行状态。需要修改一个配置文件 SPARK_HOME/slaves，添加一个 worker 节点，然后再通过 sbin 目录下的 start-all.sh 启动 Spark 集群。下面讲解完整的 Spark 伪分布式的配置。

步骤 01 配置 SSH 免密码登录。

由于启动 Spark 需要远程启动 Worker 进程，所以，需要配置从 start-all.sh 的主机到 worker 节点的 SSH 免密码登录。如果之前的章节中已经配置过此项的读者，可以不用再重复配置。

```
$ ssh-keygen -t rsa
$ ssh-copy-id server201
```

步骤 02 修改配置文件。

在 spark-conf.sh 文件中添加 JAVA_HOME 环境变量，在最前面添加即可：

```
$ vim /app/spark-3.1.1/sbin/spark-conf.sh
export JAVA_HOME=/usr/java/jdk1.8.0-281
```

修改 workers 配置文件：

```
$ vim /app/spark-3.1.1/conf/workers
server201
```

步骤 03 执行 start-all.sh 启动 Spark。

使用 start-all.sh 启动 Spark：

```
$ /app/spark-3.1.1/sbin/start-all.sh
```

启动完成以后，会有两个进程，分别是 Master 和 Worker：

```
[hadoop@server201 sbin]$ jps
2128 Worker
2228 Jps
2044 Master
```

查看启动目录，可以通过访问 8080 端口查看 Web 界面。

```
$ cat /app/spark-3.1.1/logs/spark-hadoop-org.apache.spark.deploy.master.Master-1-server201.out
    21/03/22 22:03:47 INFO Utils: Successfully started service 'sparkMaster' on port 7077.
    21/03/22 22:03:47 INFO Master: Starting Spark master at spark://server201:7077
    21/03/22 22:03:47 INFO Master: Running Spark version 3.1.1
    21/03/22 22:03:47 INFO Utils: Successfully started service 'MasterUI' on port 8080.
    21/03/22 22:03:47 INFO MasterWebUI: Bound MasterWebUI to 0.0.0.0, and started at http://server201:8080
    21/03/22 22:03:47 INFO Master: I have been elected leader! New state: ALIVE
    21/03/22 22:03:50 INFO Master: Registering worker 192.168.56.201:34907 with 2 cores, 2.7 GiB RAM
```

步骤 04 再次通过 netstat 查看端口的使用情况，你会发现一共有两个端口被占用，它们分别是 7077 和 8080。

```
[hadoop@server201 sbin]$ netstat -nap | grep java
```

```
    (Not all processes could be identified, non-owned process info
     will not be shown, you would have to be root to see it all.)
    tcp6       0      0 :::8080                 :::*              LISTEN      2044/java
    tcp6       0      0 :::8081                 :::*              LISTEN      2128/java
    tcp6       0      0 192.168.56.201:34907    :::*              LISTEN      2128/java
    tcp6       0      0 192.168.56.201:7077     :::*              LISTEN      2044/java
    tcp6       0      0 192.168.56.201:53630    192.168.56.201:7077    ESTABLISHED
2128/java
    tcp6       0      0 192.168.56.201:7077     192.168.56.201:53630   ESTABLISHED
2044/java
    unix  2    [ ]         STREAM       CONNECTED      53247    2044/java
    unix  2    [ ]         STREAM       CONNECTED      55327    2128/java
    unix  2    [ ]         STREAM       CONNECTED      54703    2044/java
    unix  2    [ ]         STREAM       CONNECTED      54699    2128/java
    [hadoop@server201 sbin]$ jps
    2128 Worker
    2243 Jps
    2044 Master
```

步骤 05 查看 8080 端口。

在浏览器地址栏上直接输入 server201:8080 查看 Spark 运行的状态，如图 11-3 所示。

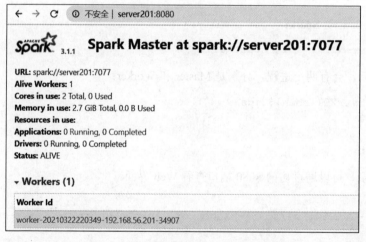

图 11-3

步骤 06 测试。

依然是使用 spark-shell，通过 --master 指定 spark://server40:7077 的地址即可使用这个集群：

```
$ spark-shell --master spark://server201:7077
```

然后再进行之前做过的 WordCount 测试。

读取文件：

```
scala> val file = sc.textFile("file:///home/hadoop/notes.txt");
file: org.apache.spark.rdd.RDD[String] = file:///home/hadoop/notes.txt MapPa
rtitionsRDD[1] at textFile at <console>:24
```

使用空格和回车键将文本分为一个一个的单词：

```
scala> val words = file.flatMap(_.split("\\s+"));
words: org.apache.spark.rdd.RDD[String] = MapPartitionsRDD[2] at flatMap at <
console>:25
```

进行统计，每一个单词统计为 1：

```
scala> val kv = words.map((_,1));
kv: org.apache.spark.rdd.RDD[(String, Int)] = MapPartitionsRDD[3] at map at <
console>:25
```

根据 key 进行统计计算：

```
scala> val worldCount = kv.reduceByKey(_+_);
worldCount: org.apache.spark.rdd.RDD[(String, Int)] = ShuffledRDD[4] at redu
ceByKey at <console>:25
```

输出并在每一行的单词和计数之间添加一个制表符号：

```
scala> worldCount.collect.foreach(kv=>println(kv._1+"\t"+kv._2));
this     2
is       1
how      1
into     2
something    1
hive.    2
file     1
And      1
process 1
you      2
about    1
wordcount        1
```

如果在运行时查看后台进程，我们将会发现以下两个进程：

```
[hadoop@server201 ~]# jps
12897 Worker
13811 SparkSubmit
13896 CoarseGrainedExecutorBackend
12825 Master
14108 Jps
```

- SparkSubmit：为一个客户端，即与 Running Application 对应。
- CoarseGrainedExecutorBackend：用于接收任务。

11.1.3 集群安装

将 Spark 目录分文件分发到其他主机，并配置 works 节点，即可快速配置 Spark 集群。请先安装好 JDK，并配置好从 Master 到 Worker 的 SSH 信任。以下是具体步骤。

步骤 01 配置计划表。

所有主机在相同目录下安装 JDK。Spark 也安装到所有主机的相同目录下，如/app/。配置信息如表 11-1 所示。

表 11-1 三台主机的配置信息

IP/主机名	软件程序	进 程
192.168.56.101 server101	JDK/Spark SSH 向 server101、server102、server102 免密码登录	Master Worker
192.168.56.102 server102	JDK/Spark	Worker
192.168.56.103 server103	JDK/Spark	Worker

步骤 02 解压并配置 Spark。

在 Server101 上解压 Spark：

```
$ tar -zxvf ~/spark-3.1.2.tar.gz -C /app/
```

修改 spark-conf.sh 文件，在最开始添加 JAVA_HOME 环境变量：

```
$ vim /app/spark-3.1.1/sbin/spark-conf.sh
export JAVA_HOME=/usr/java/jdk1.8.0-281
```

修改 worker 文件，添加所有主机名称：

```
$vim /app/spark-3.1.2/conf/workers
server101
server102
server103
```

然后使用 scp 将 Spark 目录分到所有主机相同的目录下：

```
$ scp -r /app/spark-3.1.2  server102:/app/
$ scp -r /app/spark-3.1.2  server103:/app/
```

步骤 03 启动 Spark。

最后在主 Spark 上执行 start-all.sh：

```
$ /app/spark-3.1.2/sbin/start-all.sh
```

启动完成以后，查看 master 主机的 8080 端口，如图 11-4 所示。

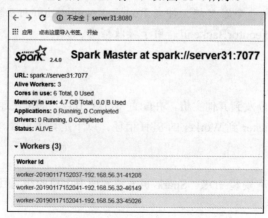

图 11-4

如果已经配置了 Hadoop 集群，且 Hadoop 的 DataNode 节点与 Spark 的 worker 节点在相同的主机上，则访问 HDFS 上的文件方式如下：

```
$spark-shell --master spark://server101:7077
scala> val rdd1 = sc.textFile( "hdfs://server101:8082/test/a.txt" );
```

将结果保存到 HDFS，最后查看 HDFS 上计算的结果即可：

```
scala> rdd1.flatMap(_.split("\\s+")).map((_,1)).reduceByKey(_+_).saveAsTextFile("hdfs://server101:8020/out004");
```

11.1.4 Spark on YARN

在 Spark Standalone 模式下，集群资源调度由 Master 节点负责。Spark 也可以将资源调度交给 YARN 来负责，其好处是 YARN 支持动态资源调度。Standalone 模式只支持简单的固定资源分配策略，每个任务固定数量的 core，各 Job 按顺序依次分配资源，资源不够时排队等待。这种策略适用单用户的场景，在多用户的情况下，各用户的程序差别很大，这种简单粗暴的策略很可能导致有些用户总是分配不到资源，而 YARN 的动态资源分配策略可以很好地解决这个问题。另外，YARN 作为通用的资源调度平台，除了为 Spark 提供调度服务外，还可以为其他子系统（比如 Hadoop MapReduce、Hive）提供调度，这样由 YARN 来统一为集群上的所有计算负载分配资源，可以避免资源分配的混乱无序。

在 Spark Standalone 集群部署完成之后，配置 Spark 支持 YARN 就相对容易多了。

步骤01 配置 spark-env.sh。

Spark 已经可以配置运行在 YARN 上，只要在 spark-env.sh 中配置 Hadoop 的配置文件信息即可。

```
$ vim /app/spark-3.1.1/conf/spark-env.sh
HADOOP_CONF_DIR=/app/hadoop-3.2.2/etc/hadoop/
YARN_CONF_DIR/app/hadoop-3.2.2/etc/hadoop/
SPARK_EXECUTOR_CORES=2
```

将 Spark 依赖的 jar 打包，上传到 HDFS 并在 spark-default.sh 中配置这个 jar 包的位置。进入 /app/spark-3.1.2/jars，然后执行如下命令：

```
$ jar -cv0f  ~/spark-libs.jar  *.jar
```

将打好的 jar 包上传到 HDFS：

```
$ hdfs dfs -put ~/spark-libs.jar /spark/jars/spark-libs.jars
```

然后在 spark-default.sh 中配置上述地址：

```
$ vim /app/spark-3.1.2/conf/spark-default.conf
spark.yarn.archive=hdfs://server201:8020/spark/jars/spark-libs.jar
```

步骤02 启动 spark-shell --master yarn。

其后就可以使用 spark-shell --master yarn 来启动 Spark 客户端。

如果你的内存不够大，在启动时，会出现以下异常：

```
  Container is running beyond virtual memory limits. Current usage: 250.2 MB of
1 GB physical memory used;
```

解决方法是取消 YARN 的内存检查。

在 yarn-site.xml 文件中添加：

```
<property>
    <name>yarn.nodemanager.vmem-check-enabled</name>
    <value>false</value>
</property>
```

配置完成以后，需要重新启动 YARN。以下是测试：

```
[hadoop@server201 /app]$ /app/spark-3.1.2/bin/spark-shell --master yarn
2021-03-17 15:49:39 WARN  NativeCodeLoader:62 - Unable to load native-hadoop
Spark context Web UI available at http://server201:4040
Spark context available as 'sc' (master = yarn, app id =
application_1547711305090_0001).
Spark session available as 'spark'.
Welcome to
      ____              __
     / __/__  ___ _____/ /__
    _\ \/ _ \/ _ `/ __/  '_/
   /___/ .__/\_,_/_/ /_/\_\   version 3.1.1
      /_/

Using Scala version 2.12.0 (Java HotSpot(TM) 64-Bit Server VM, Java 1.8.0_19
1)
Type in expressions to have them evaluated.
Type :help for more information.
```

现在做一个测试，来查看 Spark on YARN 的运行结果，首先读取 HDFS 上的一个文件。

```
scala> val rdd1 = sc.textFile("/test/a.txt");
rdd1: org.apache.spark.rdd.RDD[String] = /test/a.txt MapPartitionsRDD[1] at
textFile at <console>:24
```

统计行数：

```
scala> rdd1.count
res0: Long = 21
```

执行一系列算子：

```
scala> rdd1.flatMap(_.split("\\s+")).map((_,1)).reduceByKey(_+_).collect.for
each(kv=>println(kv._1+"\t"+kv._2));
8       9
4       1
etc     1
116     1
usr/lib 1
```

执行完成以后，查看 8088 端口和 4040 端口，如图 11-5 和图 11-6 所示。

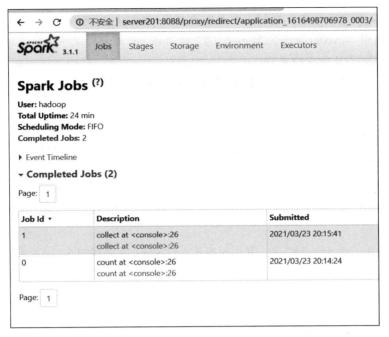

图 11-5

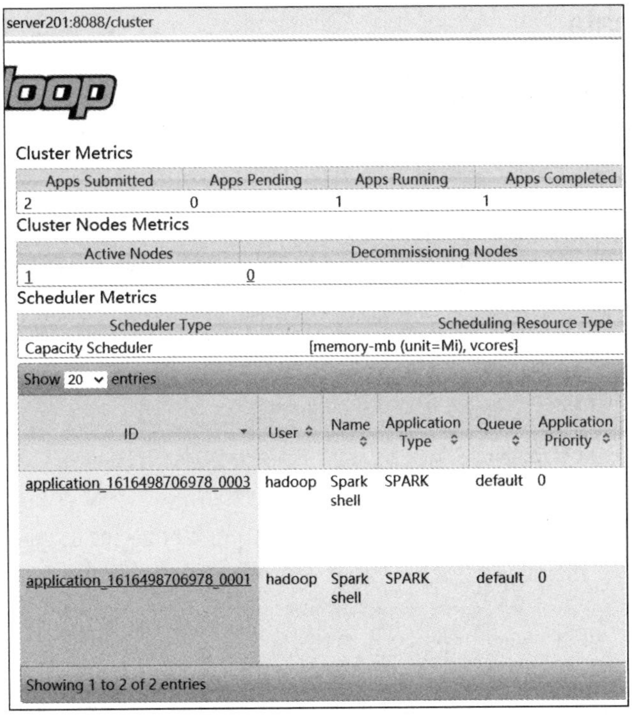

图 11-6

正如本章所讲解的，Spark 的安装与配置非常简单，它的运行方式分别是：

- 本地运行：直接使用 spark-shell --master local[*]。

- 伪分布式和完全分布式运行：首先需要使用 start-all.sh 启动 Spark 集群，然后再使用 spark-shell --master spark://server:7077 启动 driver 进程，其中 driver 进程是运行 spark 程序 main 函数所在的进程。
- 完全分布式运行在 YARN 上：这需要在配置文件 spark-env.sh 中添加 HADOOP_CONF_DIR，然后就可以使用 spark-shell --master yarn 方式来启动 driver。

11.2　使用 Scala 开发 Spark 应用

可以在 IDE 集成开发环境中，使用 Scala、Java、Python 开发 Spark 应用。本节将使用 Scala 开发 Spark 应用。

在启动 Spark 时会看到以下信息，其中说明了当前 Spark 所使用的 Scala 版本为 2.12.0。

```
Using Scala version 2.12.0 (Java HotSpot(TM) 64-Bit Server VM, Java 1.8.0_191)
Type in expressions to have them evaluated.
Type :help for more information.
```

可见 Spark 3.1.1 所使用的 Scala 是 2.12.0 版本。

11.2.1　安装 Scala

在 Windows 上使用 IDEA 安装 Scala 插件，并在本地安装 Scala 环境，与安装 JDK 环境一样，也需要配置 Scala 的环境变量。

下载 Scala 地址：

```
https://www.scala-lang.org/download/2.12.0.html
https://downloads.lightbend.com/scala/2.12.0/scala-2.12.0.zip
```

解压并配置环境变量：

```
SCALA_HOME=D:\programfiles\scala-2.12.0
PATH=%SCALA_HOME%\bin
```

登录 Scala 命令行，查看版本：

```
C:\>scala -version
Scala code runner version 2.12.0 -- Copyright 2002-2016, LAMP/EPFL and Lightbend, Inc.
```

登录 Scala 命令行：

```
D:\a>scala
Welcome to Scala 2.12.0 (Java HotSpot(TM) 64-Bit Server VM, Java 1.8.0_251).
Type in expressions for evaluation. Or try :help.
scala> 1+1
res0: Int = 2
```

在 IDEA 中安装 Scala 插件。检查已安装的 IDEA 的版本，并安装对应的 Scala 版本，如图 11-7 所示。

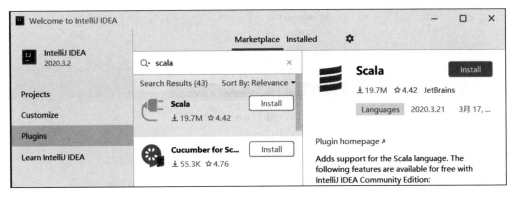

图 11-7

11.2.2 开发 Spark 程序

首先在我们已经创建的项目 chapter11 模块上，添加 Scala 支持，如图 11-8、图 11-9 所示。

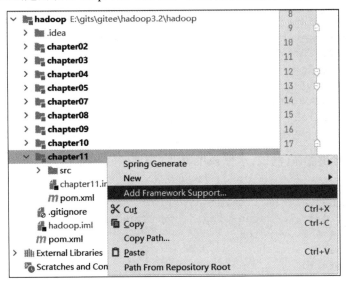

图 11-8

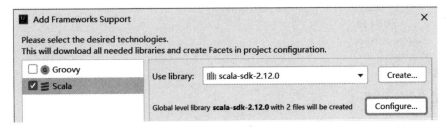

图 11-9

在 main 目录下，创建 scala 目录，并设置为 resource root，如图 11-10 所示。

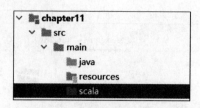

图 11-10

添加依赖:

```xml
<dependency>
    <groupId>org.apache.spark</groupId>
    <artifactId>spark-core_2.12</artifactId>
    <version>3.1.1</version>
</dependency>
```

添加编译 JDK 为 1.8 的插件:

```xml
<plugin>
    <groupId>org.apache.maven.plugins</groupId>
    <artifactId>maven-compiler-plugin</artifactId>
    <version>3.8.0</version>
    <configuration>
        <source>1.8</source>
        <target>1.8</target>
    </configuration>
</plugin>
```

为了让 Maven 能够编译、测试、运行 scala 项目,需要做如下配置(注意:不能使用 4.2.4 版本):

```xml
<plugin>
    <groupId>net.alchim31.maven</groupId>
    <artifactId>scala-maven-plugin</artifactId>
    <version>4.4.1</version>
    <executions>
        <execution>
            <goals>
                <goal>compile</goal>
                <goal>testCompile</goal>
            </goals>
        </execution>
    </executions>
</plugin>
```

1. 测试 Scala 程序

开发一个 Scala 程序,如【代码 11-1】所示。

【代码 11-1】HelloScala.scala

```scala
1.  package org.hadoop
2.  object HelloScala {
3.      def main(args: Array[String]): Unit = {
```

```
4.         println("Hello Scala")
5.     }
6. }
```

在 IDEA 中运行并输出结果：

```
Hello Scala
Process finished with exit code 0
```

可直接使用 Maven 打包，如图 11-11 所示。

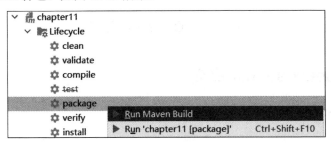

图 11-11

将 jar 文件放到任意目录下，因为需要 Scala 包的支持，使用 java-cp 运行时，必须添加 Scala 的支持包才能运行：

```
D:\a>java -cp spark-1.0.jar;%SCALA_HOME%\lib\* cn.isoft.HelloScala
Hello Scala...
```

2. 开发 Spark 程序

开发运行在本地的 Scala 程序，如【代码 11-2】所示。

【代码 11-2】WordCount.scala

```
1.  package org.hadoop.spark
2.  import org.apache.spark.rdd.RDD
3.  import org.apache.spark.{SparkConf, SparkContext}
4.  object WordCount {
5.      def main(args: Array[String]): Unit = {
6.          //声明 Config
7.          val conf: SparkConf = new SparkConf();
8.          //设置 master
9.          conf.setAppName("WordCount");
10.         conf.setMaster("local[2]");
11.         //声明 SparkContext
12.         val sc: SparkContext = new SparkContext(conf);
13.         //声明一个 RDD
14.         val rdd: RDD[String] = sc.textFile("file:///D:/a/a.txt", minPartit
ions = 2);
15.         val a: Long = rdd.count();
16.         println("size is:" + a);
17.         //以下执行一个 RDD
18.         rdd.flatMap(_.split("\\s+")).map(str => (str, 1)).reduceByKey(_ +
_) //
```

```
19.        .map(kv => kv._1 + "\t" + kv._2).saveAsTextFile("file:///D:/a/2
");
20.        sc.stop();
21.    }
22. }
```

运行上面的程序,并查看 D:\a\2 目录下输出的结果。

下一节我们将上面的程序修改成从 HDFS 上动态读取数据,并通过 spart-submit 提交程序。

11.3 spark-submit

11.3.1 使用 spark-submit 提交

首先查看 spart-submit 的帮助:

```
[hadoop@server201 app]# spark-submit
Usage: spark-submit [options] <app jar | python file | R file> [app arguments]
Usage: spark-submit --kill [submission ID] --master [spark://...]
Usage: spark-submit --status [submission ID] --master [spark://...]
Usage: spark-submit run-example [options] example-class [example args]
```

修改 Scala 程序,输入输出都通过参数来接收,如【代码 11-3】所示。

【代码 11-3】WordCount2.scala

```
1.  package org.hadoop.spark
2.  object WordCount2 {
3.     def main(args: Array[String]): Unit = {
4.        if (args.length < 2) {
5.           print("usage :<in> <out>");
6.           return;
7.        }
8.        val in: String = args.apply(0);
9.        val out: String = args.apply(1);
10.       val conf: SparkConf = new SparkConf();
11.       conf.set("fs.defaultFS", "hdfs://server201:8020");
12.       conf.setAppName("WordCount");
13.       var sc: SparkContext = new SparkContext(conf);
14.       //获取 hadoop config
15.       val hadoopConfig: Configuration = new Configuration();
16.       hadoopConfig.set("fs.defaultFS", "hdfs://server201:8020");
17.       val fs: FileSystem = FileSystem.get(hadoopConfig);
18.       val pathOut: Path = new Path(out);
19.       if (fs.exists(pathOut)) {
20.          fs.delete(pathOut, true); //删除已经存在的文件
21.       }
22.       if (!fs.exists(new Path(in))) {
23.          print("文件或目录不存在:" + in);
24.          return;
```

```
25.        }
26.        val rdd = sc.textFile(in, minPartitions = 2);
27.        rdd.flatMap(_.split("\\s+"))
28.          .map((_, 1))
29.          .reduceByKey(_ + _)
30.          .sortByKey()
31.          .map(kv => kv._1 + "\t" + kv._2)
32.          .saveAsTextFile(out);
33.        sc.stop();
34.      }
35.    }
```

使用 Maven 打包，输入命令或者使用 IDEA 打包。

上传到 Linux 然后使用以下语句提交代码：

```
$ spark-submit --master spark://server201:7077 \
--class org.hadoop.spark.WordCount2 \
chapter11-1.0.jar\
 hdfs://server201:8020/test/    \
hdfs://server201:8020/out001
```

执行完成以后，查看目录下输出的数据即可。

到目前为止，我们已经可以使用 Scala 开发 Spark 程序了。

11.3.2 spark-submit 参数说明

Spark 提交任务常见的两种模式：

- local[k]：本地使用 k 个 worker 线程运行 Saprk 程序。这种模式适合小批量数据在本地调试代码用（若使用本地的文件，则需要在前面加上 file://）。
- Spark on YARN 模式：
  - yarn-client 模式：以 client 模式连接到 YARN 集群，该方式 driver 是在 client 上运行的。
  - yarn-cluster 模式：以 cluster 模式连接到 YARN 集群，该方式 driver 运行在 worker 节点上。

对于应用场景来说，yarn-cluster 模式适合生产环境，yarn-client 模式适合交互和调试。

1. 更多通用的可选参数

（1）--master master_url：MASTER_URL，可以是 spark://host:port、mesos://host:port、yarn、yarn-cluster、yarn-client、local。

（2）--deploy-mode：DEPLOY_MODE，driver 程序运行的地方，client 或者 cluster，默认是 client。

（3）--class：CLASS_NAME，主类名称，含包名。

（4）--jars：逗号分隔的本地 JARS，driver 和 executor 依赖的第三方 jar 包。

（5）--driver-class-path：驱动依赖的 jar 包。

（6）--files：用逗号隔开的文件列表，会放置在每个 executor 工作目录中。

（7）--conf：Spark 的配置属性。例如：

- spark.executor.userClassPathFirst=true：当在 executor 中加载类时，用户添加的 jar 是否比 Spark 自己的 jar 优先级高。这个属性可以降低 Spark 依赖和用户依赖的冲突。
- spark.driver.userClassPathFirst=true：当在 driver 中加载类时，用户添加的 jar 是否比 Spark 自己的 jar 优先级高。这个属性可以降低 Spark 依赖和用户依赖的冲突。

（8）--driver-memory：driver 程序使用内存大小（例如：1000MB、5GB），默认为 1024MB。

（9）--executor-memory：每个 executor 内存大小（如：1000MB、2GB），默认 1GB。

2. 仅限于 Spark on YARN 模式

（1）--driver-cores：driver 使用的 core，仅在 cluster 模式下，默认为 1。

（2）--queue：QUEUE_NAME 指定资源队列的名称，默认为 default。

（3）--num-executors：一共启动的 executor 数量，默认为 2。

（4）--executor-cores：每个 executor 使用的内核数，默认为 1。

3. 几个重要的参数说明

（1）executor_cores*num_executors：表示的是能够并行执行 Task 的数目不宜太小或太大。一般不超过总队列 cores 的 25%，比如队列总 cores 400，最大不要超过 100，最小不低于 40，除非日志量很小。

（2）executor_cores：不宜为 1。否则 work 进程中线程数过少，一般 2~4 为宜。

（3）executor_memory：一般 6~10GB 为宜，最大不超过 20GB，否则会导致 GC 代价过高，或资源浪费严重。

（4）driver-memory：driver 不做任何计算和存储，只是下发任务与 YARN 资源管理器和 task 交互，除非是 spark-shell，否则一般为 1~2GB。

增加每个 executor 的内存量，增加了内存量以后，对性能的提升有三点：

- 如果需要对 RDD 进行 cache，那么更多的内存就可以缓存更多的数据，将更少的数据写入磁盘，甚至不写入磁盘，由此减少了磁盘 IO。
- 对于 shuffle 操作，reduce 端需要内存来存放拉取的数据并进行聚合。如果内存不够，也会写入磁盘。如果给 executor 分配更多内存以后，就有更少的数据需要写入磁盘，甚至不需要写入磁盘，由此减少了磁盘 IO，提升了性能。
- 对于 task 的执行，可能会创建很多对象。如果内存比较小，可能会频繁导致 JVM 堆内存满了，然后频繁 GC（垃圾回收），minor GC 和 full GC（速度很慢）。内存加大以后，带来更少的垃圾回收，避免了速度变慢，提升了性能。

Spark 提交参数的设置非常重要，如果设置得不合理，就会影响性能。所以大家要根据具体的情况适当地调整参数的配置，有助于提高程序执行的性能。

11.4 DataFrame

Spark 的运行和计算都慢慢转向围绕 DataFrame 来进行。DataFrame 可以看成一个简单的"数据矩阵(数据框)"或"数据表",对其进行操作也只需要调用有限的数组方法即可。它与一般"表"的区别在于:DataFrame 是分布式存储,可以更好地利用现有的云数据平台,并在内存中运行。

11.4.1 DataFrame 概述

DataFrame 实质上是存储在不同节点计算机中的一张关系型数据表。分布式存储最大的好处是:可以让数据在不同的工作节点(worker)上并行存储,以便在需要数据的时候并行运算,从而获得最迅捷的运行效率。

1. DataFrame 与 RDD 的关系

RDD(Resilient Distributed Datasets)是一种分布式弹性数据集,将数据分布存储在不同节点的计算机内存中进行存储和处理。每次 RDD 对数据处理的最终结果,都分别存放在不同的节点中。Resilient 是弹性的意思,在 Spark 中指的是数据的存储方式,即数据在节点中进行存储时候既可以使用内存也可以使用磁盘。这为使用者提供了很大的自由,提供了不同的持久化和运行方法,是一种有容错机制的特殊数据集合。

RDD 可以说是 DataFrame 的前身,DataFrame 是 RDD 的发展和拓展。RDD 中可以存储任何单机类型的数据,但是直接使用 RDD 在字段需求明显时,存在算子难以复用的缺点。例如,假设 RDD 存的数据是一个 Person 类型的数据,现在要求出所有年龄段(10 年一个年龄段)中最高的身高与最大的体重。使用 RDD 接口时,因为 RDD 不了解其中存储的数据的具体结构,需要用户自己去写一个很特殊化的聚合函数来完成这样的功能。那么如何改进才可以让 RDD 了解其中存储的数据包含哪些列并在列上进行操作呢?

根据谷歌的解释,DataFrame 是表格或二维数组状结构,其中每一列包含对一个变量的度量,每一行包含一个案例,类似于关系型数据库中的表或者 R/Python 中的 DataFrame,可以说是一个具有良好优化技术的关系表。

有了 DataFrame,框架会了解 RDD 中的数据具有什么样的结构和类型,使用者可以说清楚自己对每一列进行什么样的操作,这样就有可能实现一个算子,用在多列上比较容易进行算子的复用。甚至可以,在需要同时求出每个年龄段内不同的姓氏有多少个的时候使用 RDD 接口,在之前的函数需要很大的改动才能满足需求时使用 DataFrame 接口,这时只需要添加对这一列的处理,原来的 max/min 相关列的处理都可保持不变。

在 Apache Spark 里,DataFrame 优于 RDD,但也包含了 RDD 的特性。RDD 和 DataFrame 的共同特征是不可变性、内存运行、弹性、分布式计算能力,即 DataFrame = RDD[Row] + shcema。

分布式数据的容错性处理是涉及面较广的问题,较为常用的方法主要有两种:

- 检查节点:对每个数据节点逐个进行检测,随时查询每个节点的运行情况。这样做的好处是便于操作主节点,随时了解任务的真实数据运行情况;坏处是系统进行的是分布式存储和运算,节点检测的资源耗费非常大,而且一旦出现问题,就需要将数据在不同节点中搬运,反而更加耗费时间,也会极大地拉低执行效率。
- 更新记录:运行的主节点并不总是查询每个分节点的运行状态,而是将相同的数据在不同的节点(一般情况下是 3 个)中进行保存,各个工作节点按固定的周期更新在主节点中运行的记录,如果在一定时间内主节点查询到数据的更新状态超时或者有异常,就在存储相同数据的不同节点上重新启动数据计算工作。其缺点在于数据量过大时,更新数据和重新启动运行任务的资源耗费也相当大。

2. DataFrame 理解及其特性

DataFrame 是一个不可变的分布式数据集合,与 RDD 不同,数据被组织成命名列,就像关系型数据库中的表一样,即具有定义好的行、列的分布式数据表。

DataFrame 背后的思想是允许处理大量结构化数据。DataFrame 包含带 schema 的行。schema 是数据结构的说明,意为模式。schema 是 Spark 的 StructType 类型,由一些域(StructFields)组成,域中明确了列名、列类型以及一个布尔类型的参数(表示该列是否可以有缺失值或 null 值),最后还可以选择地明确该列关联的元数据(在机器学习库中,元数据是一种存储列信息的方式,平常很少用到)。schema 提供了详细的数据结构信息,例如包含哪些列,每列的名称和类型各是什么。DataFrame 由于其表格格式而具有其他元数据,这使得 Spark 可以在最终查询中运行某些优化。

使用一行代码即可输出 schema,代码如下:

```
df.printSchema()
//看看 schema 是什么样子
```

DataFrame 的另外一大特性是延迟计算(懒惰执行),即一个完整的 DataFrame 运行任务被分成两部分:Transformation 和 Action(转化操作和行动操作)。转化操作就是从一个 RDD 产生一个新的 RDD,行动操作就是进行实际的计算。只有当执行一个行动操作时,才会执行并返回结果。下面仍然以 RDD 这种数据集解释一下这两种操作。

(1)Transformation

Transformation 用于创建 RDD。在 Spark 中,RDD 只能使用 Transformation 创建,同时 Transformation 还提供了大量的操作方法,例如 map、filter、groupBy、join 等。除此之外,还可以利用 Transformation 生成新的 RDD,在有限的内存空间中生成尽可能多的数据对象。有一点要牢记,无论发生了多少次 Transformation,在 RDD 中真正数据计算运行的操作都不可能真正运行。

(2)Action

Action 是数据计算的执行部分,通过执行 count、reduce、collect 等方法真正执行数据的计算部分。实际上,RDD 中所有的操作都是使用 Lazy 模式(一种程序优化的特殊形式)进行的。运行在编译的过程中,不会立刻得到计算的最终结果,而是记住所有的操作步骤和方法,只有显式地遇到启动命令才进行计算。

这样做的好处在于大部分优化和前期工作在 Transformation 中已经执行完毕,当 Action 进行工作时只需要利用全部资源完成业务的核心工作即可。

Spark SQL 可以使用其他 RDD 对象、parquet 文件、json 文件、Hive 表以及通过 JDBC 连接到

其他关系型数据库作为数据源,来生成 DataFrame 对象。

11.4.2 DataFrame 基础应用

1. 创建 DataFrame

Spark 3 推荐使用 SparkSession 来创建 Spark 会话,然后使用 SparkSession 创建出来的 Application 来创建 DataFrames。

```
import org.apache.spa
val spark = SparkSession
  .builder()                                              //创建 Spark 会话
  .appName("Spark SQL basic example")                     //设置会话名称
  .master("local")                                        //设置本地模式
  .config("spark.some.config.option", "some-value")       //设置相关配置
  .getOrCreate()                                          //创建会话变量
```

对于所有的 Spark 功能,SparkSession 类都是入口,所以创建基础的 SparkSession 只需要使用 SparkSession.builder()。使用 SparkSession 时,应用程序能够从现存的 RDD、Hive Table 或者 Spark 数据源里面创建 DataFrame。

以下是完整的 createDataFrame 方法:

```
import org.apache.spark.sql._
import org.apache.spark.sql.types._
val sparkSession = new org.apache.spark.sql.SparkSession(sc)
val schema =
  StructType(
    StructField("name", StringType, false) ::
    StructField("age", IntegerType, true) :: Nil)
val people =
  sc.textFile("examples/src/main/resources/people.txt").map(
    _.split(",")).map(p => Row(p(0), p(1).trim.toInt))
val dataFrame = sparkSession.createDataFrame(people, schema)
dataFrame.printSchema
//root
//| -- name: string (nullable = false)
//| -- age: integer (nullable = true)
dataFrame.createOrReplaceTempView("people")
sparkSession.sql("select name from people").collect.foreach(println)
```

(1)第一种方式:从上面代码中可以看出,createDataFrame 方法在使用时是借助于 SparkSession 会话环境进行工作的,因此需要对 Spark 会话环境变量进行设置。以上代码先从一个文件里创建一个 RDD 再使用 createDataFrame 方法,其中第一个参数是 RDD、第二个参数 schema 是上面定义的 DataFrame 的字段数据类型等信息。

(2)第二种方式:使用 toDF()函数将此强类型数据集合转换为带有重命名列的通用 DataFrame。这在将元组的 RDD 转换为具有有意义名称的 DataFrame 时非常方便。

```
import spark.implicits._
```

```
val rdd: RDD[(Int, String)] = ...
rdd.toDF()        //这里是一个隐式转换，将 RDD 变成列名为_1、_2 的 DataFrame
rdd.toDF("id", "name")      //将 RDD 变成列名为"id"、"name"的 DataFrame，
//需要添加结构信息并加上列名
```

提示：对 DataFrame、DataSet 和 RDD 进行转换需要 import spark.implicits._ 这个包的支持。

2. select 和 selectExpr 方法

select 和 selectExpr 方法用于把 DataFrame 中的某些列筛选出来。其中，select 用来选择某些列出现在结果集中，结果作为一个新的 DataFrame 返回，使用方法如下：

```
import org.apache.spark.sql.SparkSession
object select {
   def main(args: Array[String]): Unit = {
      val spark = SparkSession
        .builder()                                //创建 Spark 会话
        .appName("Spark SQL basic example")       //设置会话名称
        .master("local")                          //设置本地模式
        .getOrCreate()                            //创建会话变量
      val rdd = spark.sparkContext.parallelize(Array(1,2,3,4))
      import spark.implicits._
      val df = rdd.toDF("id")
      df.select("id").show()                      //选择"id"列
   }
}
```

打印结果如下：

```
+---+
| id|
+---+
|  1|
|  2|
|  3|
|  4|
+---+
```

如果是 selectExpr 方法，则代码如下：

```
import org.apache.spark.sql.SparkSession
object select {
   def main(args: Array[String]): Unit = {
      val spark = SparkSession
        .builder()                                //创建 Spark 会话
        .appName("Spark SQL basic example")       //设置会话名称
        .master("local")                          //设置本地模式
        .getOrCreate()                            //创建会话变量
      val rdd = spark.sparkContext.parallelize(Array(1,2,3,4))
      import spark.implicits._
      val df = rdd.toDF("id")
      df.selectExpr("id as ID").show()            //设置了一个别名 ID
```

```
    }
}
```

具体结果请读者自行运行查看。

3. collect 方法

collect 方法将已经存储的 DataFrame 数据从存储器中收集回来，并返回一个数组，包括 DataFrame 集合所有的行，其源码如下：

```
def collect(): Array[T]
```

Spark 的数据是分布式存储在集群上的，如果想获取一些数据在本机 Local 模式上操作，就需要将数据收集到 driver 驱动器中。collect()返回 DataFrame 中的全部数据，并返回一个 Array 对象，代码如下：

```
import org.apache.spark.sql.SparkSession
object collect {
    def main(args: Array[String]): Unit = {
        val spark = SparkSession
            .builder()                                       //创建 Spark 会话
            .appName("Spark SQL basic example")              //设置会话名称
            .master("local")                                 //设置本地模式
            .getOrCreate()                                   //创建会话变量
        val rdd = spark.sparkContext.parallelize(Array(1,2,3,4))
        import spark.implicits._
        val df = rdd.toDF("id")
        val arr = df.collect()
        println(arr.mkString("Array(", ", ", ")"))
    }
}
```

注意：将数据收集到驱动器中，尤其是当数据集很大或者分区数据集很大时，很容易让驱动器崩溃。数据收集到驱动器中进行计算，就不是分布式并行计算了，而是串行计算，会更慢。所以，除了查看小数据，一般不建议使用。

4. DataFrame 计算行数 count 方法

count 方法用来计算数据集 DataFrame 中行的个数，返回 DataFrame 集合的行数，使用方法如下：

```
import org.apache.spark.sql.SparkSession
object count {
    def main(args: Array[String]): Unit = {
        val spark = SparkSession
            .builder()                                       //创建 Spark 会话
            .appName("Spark SQL basic example")              //设置会话名称
            .master("local")                                 //设置本地模式
            .getOrCreate()                                   //创建会话变量
        val rdd = spark.sparkContext.parallelize(Array(1,2,3,4))
        import spark.implicits._
        val df = rdd.toDF("id")
        println(df.count())                                  //计算行数
```

```
        }
}
```

最终结果如下:

```
4
```

5. 过滤数据的 filter 方法

filter 方法是一个比较常用的方法,用来按照条件过滤数据集。如果想选择 DataFrame 中某列为大于或小于某数据,就可以使用 filter 方法。对于多个条件,可以将 filter 方法写在一起。

filter 方法接收任意一个函数作为过滤条件。行过滤的逻辑是先创建一个判断条件表达式,根据表达式生成 true 或 false,然后过滤使表达式值为 false 的行。filter 方法的具体使用示例程序如下:

```
import org.apache.spark.sql.SparkSession
object fliter {
    def main(args: Array[String]): Unit = {
        val spark = SparkSession
          .builder()                                    //创建 Spark 会话
          .appName("Spark SQL basic example")           //设置会话名称
          .master("local")                              //设置本地模式
        .getOrCreate()                                  //创建会话变量
        val rdd = spark.sparkContext.parallelize(Array(1,2,3,4))
        import spark.implicits._
        val df = rdd.toDF("id")
        val df2 = df.filter("id>3")//过滤 id 列大于 3 的数据(行)或 _ >= 3
        println(df2.cache().show())                     //打印结果
    }
}
```

具体结果请读者自行验证。这里需要说明的是,"_ >= 3"采用的是 Scala 编程中的编程规范,_的作用是作为占位符标记所有传过来的数据。在此方法中,数组的数据(1,2,3,4)依次传进来替代了占位符。

6. 以整体数据为单位操作数据的 flatMap 方法

flatMap 方法是对 DataFrame 中的数据集进行整体操作的一个特殊方法,因为其在定义时是针对数据集进行操作的,因此最终返回的也是一个数据集。flatMap 方法首先将函数应用于此数据集的所有元素,然后将结果展平,从而返回一个新的数据集。应用程序如下:

```
import org.apache.spark.sql.SparkSession
object flatmap {
    def main(args: Array[String]): Unit = {
        val spark = SparkSession
          .builder()                                    //创建 Spark 会话
          .appName("Spark SQL basic example")           //设置会话名称
          .master("local")                              //设置本地模式
          .getOrCreate()                                //创建会话变量
        val rdd = spark.sparkContext.parallelize(Seq("hello!spark", "hello!hadoop"))
        import spark.implicits._
```

```
        val df = rdd.toDF("id")
        val x = df.flatMap(x => x.toString().split("!")).collect()
        println(x.mkString("Array(", ", ", ")"))
    }
}
```

7. 分组数据的 groupBy 和 agg 方法

groupBy 方法是将传入的数据进行分组，依据是作为参数传入的计算方法。聚合操作调用的是 agg 方法，该方法有多种调用方式，一般与 groupBy 方法配合使用。在使用 groupBy 时，一般都是先分组再使用 agg 等聚合函数来对数据进行聚合。groupBy+agg 的使用方法如下：

```
import org.apache.spark.sql.SparkSession
object groupBy {
    def main(args: Array[String]): Unit = {
        val spark = SparkSession
          .builder()                                        //创建 Spark 会话
          .appName("Spark SQL basic example")               //设置会话名称
          .master("local")                                  //设置本地模式
          .getOrCreate()                                    //创建会话变量
        val df = spark.read.json("./src/C03/people.json")
        df.groupBy("name").agg("age" -> "count").show()
    }
}
```

这里采用 groupBy+agg 的方法统计了 age 字段的条数。

在 GroupedData 的 API 中提供了 groupBy 之后的操作，比如：

- max(colNames: String*)方法：获取分组中指定字段或者所有的数字类型字段的最大值，只能作用于数字型字段。
- min(colNames: String*)方法：获取分组中指定字段或者所有的数字类型字段的最小值，只能作用于数字型字段。
- mean(colNames: String*)方法：获取分组中指定字段或者所有的数字类型字段的平均值，只能作用于数字型字段。
- sum(colNames: String*)方法：获取分组中指定字段或者所有的数字类型字段的和值，只能作用于数字型字段。
- count 方法：获取分组中的元素个数。

这些都等同于 agg 方法。

8. 删除数据集中某列的 drop 方法

drop 方法从数据集中删除某列，然后返回 DataFrame 类型，使用方法的代码示例如下：

```
import org.apache.spark.sql.SparkSession
object drop {
    def main(args: Array[String]): Unit = {
        val spark = SparkSession
          .builder()                                        //创建 Spark 会话
```

```
            .appName("Spark SQL basic example")      //设置会话名称
            .master("local")                         //设置本地模式
            .getOrCreate()                           //创建会话变量
        val df = spark.read.json("./src/C03/people.json")
        df.drop("age").show()                        //删除 age 列
    }
}
```

最终打印结果如下：

```
+-------+
|   name|
+-------+
|Michael|
|   Andy|
| Justin|
+-------+
```

这里也可以用通过 select 方法来实现列的删除，不过建议使用专门的 drop 方法来实现——规范又显而易见，对于维护工作来更有效率。

11.5 Spark SQL

Spark SQL 是 Spark 的一个模块，主入口是 SparkSession，将 SQL 查询与 Spark 程序无缝混合。DataFrame 和 SQL 提供了访问各种数据源（通过 JDBC 或 ODBC 连接）的常用方法，包括 Hive、Avro、Parquet、ORC、JSON 和 JDBC。甚至可借助于 Spark SQL 实现多数据源之间的交互，因为任何数据源都可以用相同方式注册为 Spark 临时表。Spark SQL 还支持 HiveQL 语法以及 Hive SerDes 和 UDF，允许用户访问现有的 Hive 数据仓库。

Spark SQL 包括基于成本的优化器、列式存储和代码生成，以快速进行查询。同时，它使用 Spark 引擎扩展到数千个节点，该引擎提供完整的中间查询容错。不要担心使用不同的引擎来获取历史数据。

Spark 2.0 版本前，Spark SQL 发生了部分变化，Spark 2.0 版本以后，我们操作的 SQL 的对象为 DataSet，并且将把 SqlContext 和 HiveContext 整合到一起成为 SparkSession，具体如下：

Spark 2.0 之前：

- SQL 入口：SqlContext 和 HiveContext。
- SqlContext：主要 DataFrame 的构建以及 DataFrame 的执行，SqlContext 指的是 Spark 中 SQL 模块的程序入口。
- HiveContext：是 SqlContext 的子类，专门用于与 Hive 集成，比如读取 Hive 的元数据、数据存储到 Hive 表、Hive 的窗口分析函数等。

Spark 2.0 之后：

- SQL 入口：SparkSession（Spark 应用程序的一个整体入口），合并了 SqlContext 和 HiveContext。

- Spark SQL 核心抽象：DataFrame/Dataset。

11.5.1 快速示例

本小节通过多个示例，演示将 RDD 转换成 DataFrame。

创建一个 RDD：

```
scala> val rdd=sc.makeRDD(Seq(("Jack",24),("Mary",34)));
```

转成 DataFrame：

```
scala> val df1 = rdd.toDF();
df1: org.apache.spark.sql.DataFrame = [_1: string, _2: int]
```

使用 show 显示数据：

```
scala> df1.show();
+----+---+
|  _1| _2|
+----+---+
|Jack| 24|
|Mary| 34|
+----+---+
```

1. 给 DataFrame 设置别名

```
scala> val df2 = rdd.toDF("name","age");
```

再次使用 show 显示时，将显示列的名称：

```
scala> df2.show();
+----+---+
|name|age|
+----+---+
|Jack| 24|
|Mary| 34|
+----+---+
```

2. 使用 SqlContext 执行 SQL 语句

先保存成 view：

```
scala> df2.createOrReplaceTempView("person");
```

声明 SqlContext 对象：

```
scala> val sqlContext = spark.sqlContext
```

执行 SQL 语句：

```
scala> sqlContext.sql("select * from person").show();
+----+---+
|name|age|
+----+---+
|Jack| 24|
```

```
|Mary| 34|
+----+---+
```

3. Scala 代码将 RDD 转成 Bean

在 Scala 项目中使用，需要添加 spark-sql 的依赖：

```xml
<dependency>
    <groupId>org.apache.spark</groupId>
    <artifactId>spark-sql_2.12</artifactId>
    <version>3.1.1</version>
</dependency>
```

准备一个文本文件，一行为一个对象，每一个值之间用空格分开，文件名 D:/a/stud.txt 内容如下：

```
4 Jack 34 男
1 Mike 23 女
2 刘长友 45 男
3 雪丽 27 女
```

编写以下代码，读取 stud.txt 文件的内容，并封装到 class Stud 对象中。

【代码 11-4】RddToBean.scals

```scala
1. package org.hadoop.spark
2. object RddToBean {
3.     def main(args: Array[String]): Unit = {
4.         val conf: SparkConf = new SparkConf();
5.         conf.setMaster("local[*]");
6.         conf.setAppName("RDDToBean");
7.         val spark: SparkSession = SparkSession.builder().config(conf).getOrCreate();
8.         val sc: SparkContext = spark.sparkContext;
9.         //读取文件
10.         import spark.implicits._;
11.         //2：读取文件
12.         val rdd: RDD[String] = sc.textFile("file:///D:/a/stud.txt");
13.         //3：第一次使用 map 对每一行进行 split，第二次使用 map 将数据封装到 Bean 中，最后使用 toDF 转换成 DataFrame
14.         val df = rdd.map(_.split("\\s+")).map(arr => {
15.             Stud(arr(0).toInt, arr(1), arr(2).toInt, arr(3));
16.         }).toDF();
17.         //4：显示或是保存数据
18.         df.show();
19.         spark.close();
20.     }
21.     /** 1: 声明 JavaBean，并直接声明主构造方法 **/
22.     case class Stud(id: Int, name: String, age: Int, sex: String) {
23.         /** 声明无参数的构造，调用主构造方法 **/
24.         def this() = this(-1, null, -1, null);
25.     }
26. }
```

直接在 IDEA 中运行，输出结果如下：

```
+---+------+---+---+
| id|  name|age|sex|
+---+------+---+---+
|  4|  Jack| 34| 男|
|  1|  Mike| 23| 女|
|  2| 刘长友| 45| 男|
|  3|  雪丽| 27| 女|
+---+------+---+---+
```

4. WordCount 示例

让我们再来做一次 WordCount 示例，本次使用 Spark SQL。

首先读取文件 1.txt，1.txt 中可以是任意的内容：

```
scala> val rdd = sc.textFile("file:///home/hadoop/1.txt");
```

以 \s 分为分割，并转成 DF 对象，注意转换后的对象只有一个字段 str：

```
scala> val df3 = rdd.flatMap(_.split("\\s+")).toDF("str");
```

现在可以直接使用 groupBy 进行 count 计算：

```
scala> df3.groupBy("str").count().show();
+-----+-----+
|  str|count|
+-----+-----+
|    A|    3|
|    B|    1|
|    C|    1|
...
```

还可以指定排序规则：

```
scala> df3.groupBy("str").count().sort("str").show();
+-----+-----+
|  str|count|
+-----+-----+
|   ->|    4|
|    0|    2|
...
```

或是直接使用 SQL 语句：

```
scala> df3.createOrReplaceTempView("words");
scala> sqlContext.sql("select count(str),str from words group by str").show();
+----------+-----+
|count(str)|  str|
+----------+-----+
|         3|    7|
|         1|  lib|
...
```

5. Scala 代码示例

开发 Scala 代码实现统计功能，如【代码 11-5】所示。

【代码 11-5】SparkSQL.scala

```
1.  package org.hadoop.spark
2.  object SparkSQL {
3.      def main(args: Array[String]): Unit = {
4.          val conf: SparkConf = new SparkConf();
5.          conf.setMaster("local[2]");
6.          conf.setAppName("SQL");
7.          val session: SparkSession = SparkSession.builder().config(conf).getOrCreate();
8.          val sqlContext: SQLContext = session.sqlContext;
9.          val ctx: SparkContext = session.sparkContext;
10.         val rdd: RDD[String] = ctx.textFile("file:///D:/a/stud.txt");
11.         //注意要做隐式导入
12.         import session.implicits._;
13.         val df: DataFrame = rdd.flatMap(_.split("\\s+")).toDF("str");
14.         df.show();
15.         df.groupBy("str").count().sort("str").show();
16.         //创建 View
17.         df.createTempView("words");
18.         sqlContext.sql("select str,count(str) cnt from words group by str order by str") //执行 SQL
19.             //转成 RDD
20.             .rdd
21.             .saveAsTextFile("file:///D:/a/out001"); //保存到指定目录
22.         session.close();
23.     }
24. }
```

以上程序为 Local 模式，即程序在本地运行，主要用于调试程序方便。本地测试成功后再打包在集群上运行。集群模式需要对代码进行如下修改。

【代码 11-6】SparkSQL2.scala

```
1.  package org.hadoop.spark
2.  object SparkSQL2 {
3.      def main(args: Array[String]): Unit = {
4.          if (args.length != 2) {
5.              println("usage : in out");
6.              return;
7.          }
8.          val inPath: String = args.apply(0);
9.          val outPath: String = args.apply(1);
10.         val hConf: Configuration = new Configuration();
11.         val fs: FileSystem = FileSystem.get(hConf);
12.         val dest: Path = new Path(outPath);
13.         if (fs.exists(dest)) {
14.             fs.delete(dest, true);
```

```
15.         }
16.         val conf: SparkConf = new SparkConf();
17.         conf.setAppName("SQL");
18.         val session: SparkSession = SparkSession.builder().config(conf).ge
tOrCreate();
19.         val sqlContext: SQLContext = session.sqlContext;
20.         val ctx: SparkContext = session.sparkContext;
21.         val rdd: RDD[String] = ctx.textFile(inPath);
22.         //注意要做隐式导入
23.         import session.implicits._;
24.         val df: DataFrame = rdd.flatMap(_.split("\\s+")).toDF("str");
25.         df.show();
26.         df.groupBy("str").count().sort("str").show();
27.         df.createTempView("words"); //创建 View
28.         sqlContext.sql("select str,count(str) cnt from words group by str
 order by str") //执行 SQL
29.             .rdd //转成 RDD
30.             .saveAsTextFile(outPath); //保存到指定目录
31.         session.close();
32.     }
33. }
```

使用 spark-submit 提交任务：

```
# spark-submit --master spark://server201:7077 --class org.hadoop.spark.Spar
kSQL2 chapter11-1.0.jar hdfs://server201:8020/test/ hdfs://server201:8020/out002
```

11.5.2　Read 和 Write

DataFrame/DataSet 提供了 Read 和 Write 方法，其中 Read 可以读取指定格式的数据，而 Write 可以写出指定格式的数据。

1. read 方法

read 方法可以读取指定格式的数据文件，官方数据格式参考地址为 http://spark.apache.org/docs/latest/sql-data-sources-load-save-functions.html。

这些数据格式包含：

```
csv     format   jdbc    json    load    option    options    orc    parquet    schema
table   text     textFile
```

在没有设置 spark.sql.sources.default 的情况下，默认读取的是 parquet 格式的文件。由于 Hive 在创建表时，可以通过 stored as parquet 存储数据为 parquet 格式，所以可以读取 Hive 的表数据查看。

首先，在 Hive 上创建一个 parquet 存储格式的表，使用 insert 写入一些记录：

```
Hive> create table teacher (id integer,name string ) stored as parquet;
Hive> insert into teacher values(1,'Mary');
Hive> insert into teacher values(2,'刘长友');
Hive> insert into teacher values(3,'Alex');
```

然后，就可以直接使用 Spark 读取 parquet 格式的数据了。

首先使用 read 来读取：

```
scala> spark.read.load("hdfs://server201:8020/user/hive/warehouse/one.db/teacher/*");
res5: org.apache.spark.sql.DataFrame = [id: int, name: string]
```

然后就可以直接显示结果：

```
scala> res5.show();
+---+-------+
| id|   name|
+---+-------+
|  2| 刘长友|
|  1|   Mary|
|  3|   Alex|
+---+-------+
```

textFile 用于读取文本文件，它会将整个文件所有行作为一个字段来处理：

```
scala> spark.read.textFile("hdfs://server201:8020/user/hive/warehouse/one.db/stud/stud.txt");
res7: org.apache.spark.sql.Dataset[String] = [value: string]
scala> res7.show();
+---------+
|    value|
+---------+
|   1 Jack|
...
```

2. write 方法

DataFrame/Dataset 的 write.save 方法，默认将数据保存成 parquet 形式。

下面示例是将一个 dataFrame 的数据保存到 HDFS 上。

首先声明：

```
scala> val rdd = sc.parallelize(Seq((1,"Jack"),(2,"Mary")));
```

转成 DataFrame：

```
scala> val df = rdd.toDF("id","name");
```

保存：

```
scala> df.write.save("hdfs://server201:8020/out001");
```

保存成 json 格式：

```
scala> df2.write.format("json").save("hdfs://server21:8020/out003");
```

11.6　Spark Streaming

Spark Streaming 是 Spark API 核心的扩展，支持可扩展、高吞吐量、实时数据流的容错流处理。

数据可以从 Kafka、Flume、Kinesis 或 TCP Socket 等许多来源摄入，并且可以使用与像高级别功能表达复杂的算法来处理 map、reduce、join 和 window。最后，处理的数据可以推送到文件系统，数据库和实时仪表板（DashBorad）。事实上，你可以将 Spark 的机器学习（ML）和图形处理（Graphx）算法应用于数据流。图 11-12 所示为官方图例。

图 11-12

Spark Streaming 提供了一个高层次的抽象，称为离散流 DStream，它代表连续的数据流。DStream 可以通过 Kafka、Flume 和 Kinesis 等来源的输入数据流创建，也可以通过在其他 DStream 上应用高级操作来创建。在内部表示为一系列的 RDD。

11.6.1 快速示例

根据官网的示例，快速开始一个 Spark Streaming 的使用示例。此程序将监听某个端口的数据，并将接收到的数据输出到控制台。开发 Spark Streaming 程序，需要添加 Spark Streaming 的依赖如下：

```
<dependency>
    <groupId>org.apache.spark</groupId>
    <artifactId>spark-streaming_2.12</artifactId>
    <version>3.1.1</version>
</dependency>
```

步骤 01 开发 Scala 程序。

【代码 11-7】Streaming.java

```
1.  package org.hadoop.spark
2.  object Streaming {
3.    def main(args: Array[String]): Unit = {
4.      val conf = new SparkConf()
5.        .setMaster("local[2]") //至少两个线程
6.        .setAppName("NetworkWordCount")
7.      //声明 Spark Streaming 对象
8.      val ssc = new StreamingContext(conf, Seconds(2))
9.      //声明监听的服务器及端口
10.     val lines = ssc.socketTextStream("192.168.56.201", 9999)
11.     //输出接收到的这个端口的数据
12.     lines.print();
13.     //开始运行
14.     ssc.start();
15.     ssc.awaitTermination();
16.   }
17. }
```

步骤 02 向 9999 端口发送数据。

首先在 Linux 上安装 nc 软件：

```
$ sudo yum install -y nc
```

启动 9999 端口：

```
$ nc -lk 9999
```

并向此端口输入一些数据：

```
Jack Mary
```

步骤 03 启动 Scala 程序并查看输出。

程序启动以后，每 2s 会输出一次访问数据，如果访问端口时获取到数据，就会输出数据到控制台，结果如下所示：

```
-------------------------------------------
Time: 1547961188000 ms
-------------------------------------------
Jack Mary
```

也可以将 lines.print() 修改成以下代码，即对数据进行处理以计算 WordCount：

```
lines.flatMap(_.split("\\s+")).map((_, 1)).reduceByKey(_ + _).print();
```

步骤 04 打包运行。

打包以后，通过 spark-submit 提交运行，可以得到相同的效果：

```
# spark-submit --master spark://server201:7077 \
--class org.hadoop.spark.Streaming chapter11-1.0.jar
```

步骤 05 打包运行在 Spark 集群上。

使用 spark-submit 提交的代码，如果 Spark 集群中仅有一个 Worker，则 Spark Streaming 无法运行。所以，必须保证 Worker 节点>=2 个。

修改上面的代码，删除 conf.setMaster("local[2]") 语句，删除后，配置将默认为 conf.setMaster("spark://server101:7077")，完整代码如下：

【代码 11-8】Streaming2.scala

```
1.   package org.hadoop.spark
2.   import org.apache.spark.SparkConf
3.   import org.apache.spark.streaming.{Seconds, StreamingContext}
4.   object Streaming2 {
5.     def main(args: Array[String]): Unit = {
6.       val conf: SparkConf = new SparkConf();
7.       conf.set("fs.defaultFS", "hdfs://server201:8020");
8.       conf.setAppName("Streaming01");
9.       val streamingContext: StreamingContext = new StreamingContext(conf, Seconds(5));
10.      var lines = streamingContext.socketTextStream("server101", 4444);
```

```
11.         lines.print();
12.         lines.saveAsTextFiles("hdfs://server201:8020/out001/a");
13.         streamingContext.start();
14.         streamingContext.awaitTermination();
15.     }
16. }
```

启动 Spark 集群，查看 Worker 个数如图 11-13 所示。

图 11-13

先启动 nc 命令：

```
$ nc -lk 4444
```

打包并启动：

```
$ spark-submit --master spark://server201:7077 \
--class org.hadoop.spark.Streaming2 chapter11-1.0.jar
```

启动后，可以看到 Cores 的个数为 2，因为有两个 Worker，所以 Cores 为 2，如图 11-14 所示。

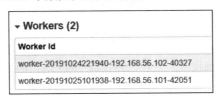

图 11-14

输入并接收数据，查看输出结果：

```
-------------------------------------------
Time: 1571972220000 ms
-------------------------------------------
Rose
Hello
```

查看 HDFS 上的数据，如图 11-15 所示。

图 11-15

11.6.2 DStream

DStream 是 Spark Streaming 的编程模型，DStream 的操作包括输入、转换和输出。图 11-16 所示是官方给出的 DStream 的图例。

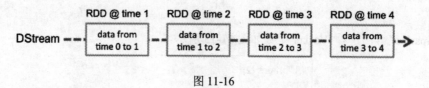

图 11-16

以下是一个 DStream 的示例程序，一样也是读取 9999 端口的数据并输出到控制台。

【代码 11-9】DStreaming.scala

```
1.   package org.hadoop.spark
2.   object DStreaming {
3.     def main(args: Array[String]): Unit = {
4.       //配置 SparkConf 参数
5.       val sparkConf: SparkConf = new SparkConf()
6.         .setAppName("SparkStreamingTCP").setMaster("local[2]")
7.       //构建 SparkContext 对象
8.       val sc: SparkContext = new SparkContext(sparkConf)
9.       //设置日志输出级别
10.      sc.setLogLevel("WARN")
11.      //构建 StreamingContext 对象，每个批处理的时间间隔
12.      val scc: StreamingContext = new StreamingContext(sc, Seconds(5))
13.      //注册一个监听的 IP 地址和端口，用来收集数据
14.      val lines: ReceiverInputDStream[String] =
15.        scc.socketTextStream("192.168.56.201", 9999)
16.      //切分每一行记录
17.      val words: DStream[String] = lines.flatMap(_.split(" "))
18.      //每个单词记为 1
19.      val wordAndOne: DStream[(String, Int)] = words.map((_, 1))
20.      //分组聚合
21.      val result: DStream[(String, Int)] = wordAndOne.reduceByKey(_ + _)
22.      //打印数据
23.      result.print()
24.      scc.start()
25.      scc.awaitTermination()
26.    }
27.  }
```

11.6.3 FileStream

FileStream 更准确地说应该叫目录文件流，它的功能是用于监听一个目录下的文件增加，如果增加了一个文件，将会读取这个文件的内容，并做处理。注意，此类在 Windows 环境下无效，只能运行在 Unix/Linux/Mac 环境下。

1. 监听 Linux 上的某个目录

【代码 11-10】FileStreaming.scala

```
1.  package org.hadoop.spark
2.  object FIleStreaming {
3.    def main(args: Array[String]): Unit = {
4.      val conf = new SparkConf()
5.        .setMaster("local[2]") //至少两个线程
6.        .setAppName("FileStream")
7.      //声明 Spark Streaming 对象
8.      val ssc = new StreamingContext(conf, Seconds(2));
9.      val context = ssc.sparkContext
10.     context.setLogLevel("WARN");
11.     //监听这个目录下文件的增加,如果有文件增加,将会读取文件内容,并处理
12.     val rdd: DStream[String] = ssc.textFileStream("file:///home/hadoop/1");
13.     rdd.flatMap(_.split("\\s+")).map((_, 1)).reduceByKey(_ + _).print();
14.     ssc.start();
15.     ssc.awaitTermination();
16.     Thread.sleep(1000*20);
17.     ssc.stop();
18.   }
19. }
```

只能在 Linux 上的获取目录中文件的增加,在 Windows 上测试不能通过。

2. 监听 HDFS 上某个目录下文件的增加

如果监听 HDFS 上某个目录下出现以下异常:

```
2019-01-20 15:32:20 ERROR JobScheduler:91 - Error generating jobs for time 1547969540000 ms
org.apache.hadoop.mapreduce.lib.input.InvalidInputException: Input path does not exist: hdfs://server21:8020/test/1.txt._COPYING_
    at org.apache.hadoop.mapreduce.lib.input.FileInputFormat.singleThreadedListStatus(FileInputFormat.java:325)
    at org.apache.hadoop.mapreduce.lib.input.FileInputFormat.listStatus(FileInputFormat.java:265)
```

其原因是,使用 HDFS 上传文件时,会先形成一个_COPYING_的文件,文件上传完成以后会被删除,但这个_COPYING_的文件也会被 Spark Stream 监听,当处理时已经被删除,所以出现上述错误。此时,就必须使用 FileStream,因为它拥有第二个参数 filter。

修改【代码 11-10】,注意使用 FileStream 而不是 textFileStream。

【代码 11-11】FileStreaming2.scala

```
1.  package org.hadoop.spark
2.  object FIleStreaming2 {
3.    def main(args: Array[String]): Unit = {
4.      if (args.length != 1) {
5.        println("usage : <in>")
6.        return;
7.      }
8.      val conf = new SparkConf()
9.        .setAppName("FileStream")
10.     //声明 Spark Streaming 对象
11.     val ssc = new StreamingContext(conf, Seconds(2));
12.     val rdd: DStream[(LongWritable, Text)] =
```

```
13.         ssc.fileStream[LongWritable, Text, TextInputFormat](directory = ar
gs(0), //目录
14.         filter = ((path: Path) => (!path.getName.contains("_COPYING_"))),
//过滤条件
15.         newFilesOnly = true);
16.     rdd.map(_._2.toString). //先获取 Value 数据
17.      flatMap(_.split("\\s+")) //再对文本进行处理
18.      .map((_, 1)).reduceByKey(_ + _).print();
19.    ssc.start();
20.    ssc.awaitTermination();
21.    Thread.sleep(1000 * 20);
22.    ssc.stop();
23.   }
24. }
```

然后打包,再执行 spark-submit 命令提交代码:

```
$spark-submit --master spark://server201:7077 \
--class org.hadoop.spark.FileStreaming2 \
chapter11-1.0.jar hdfs://server201:8020/test/
```

使用 HDFS 上传到 /test 目录下文件,查看 Steaming 方面输出的信息即可。

11.6.4　窗口函数

window 窗口函数可以计算连续的任何多个 DStream 中 RDD 的数据。如图 11-17 所示。

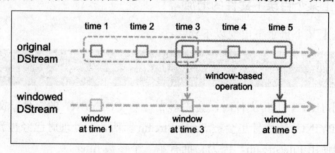

图 11-17

两个主要的参数介绍如下:

- window length: 窗口的个数。
- sliding interval: 滑动的块数。

为了处理窗口程序,PairDStreamFunctions 提供了一个函数 reduceByKeyAndWindow,该函数将在本节案例代码中使用,这个函数的源码为:

```
def reduceByKeyAndWindow(
    reduceFunc: (V, V) => V,
    windowDuration: Duration,
    slideDuration: Duration
 ): DStream[(K, V)] = ssc.withScope {
    reduceByKeyAndWindow(reduceFunc, windowDuration, slideDuration, default
```

```
Partitioner())
    }
```

现在我们开发一个 Java 客户端,每 2 秒向 9999 端口请求一次数据,设置窗口个数为 3,即 2×3=6,设置滑块个数为 2,即 2×2=4。现在我们在 nc 控制台输入数据并查看 Streaming 接收到的数据,如【代码 11-12】所示。

【代码 11-12】WindowFun.scala

```scala
package org.hadoop.spark
object WindowFun {
    def main(args: Array[String]): Unit = {
        val conf = new SparkConf()
          .setMaster("local[2]")
          .setAppName("WindowFun")
        //声明 Spark Streaming 对象
        val ssc = new StreamingContext(conf, Seconds(2));
        val context = ssc.sparkContext;
        context.setLogLevel("WARN");
        //必须声明检查点
        ssc.checkpoint("file:///D:/a/log");
        val rdd = ssc.socketTextStream("server201", 9999, StorageLevel.MEMORY_AND_DISK);
        val rdd2 =  rdd.flatMap(_.split("\\s+")).map((_,1))
          .reduceByKey(_+_).reduceByKeyAndWindow((a:Int,b:Int)=>a+b,Seconds(6),Seconds(4));
        rdd2.print();
        //开始
        ssc.start();
        ssc.awaitTermination();
    }
}
```

11.6.5 updateStateByKey

updateStateByKey 用于计算从向 Spark Streaming 传递数据开始到目前接收到的所有数据的统计信息。

为了可以均匀地向 Spark Streaming 传递 socket 数据,这里使用 Java 代码发送。

【代码 11-13】ForUpdateStateByKey.java

```java
1. package org.hadoop.spark;
2. public class ForUpdateStateByKey {
3.     public static void main(String[] args) throws Exception {
4.         ServerSocket ss = new ServerSocket(9999);
5.         while (true) {
6.             System.out.println("等待连接...");
7.             Socket client = ss.accept();
8.             System.out.println("连接,启动线程..." + client);
9.             new MyThread(client).start();
10.        }
```

```
11.         }
12.         /**
13.          * 使用以下线程，向连接成功的客户端发送数据，第一秒发送一次相同的数据
14.          **/
15.         public static class MyThread extends Thread {
16.             private Socket client;
17.             public MyThread(Socket client) {
18.                 this.client = client;
19.             }
20.             @Override
21.             public void run() {
22.                 try {
23.                     int times = 1;
24.                     PrintStream ps = new PrintStream(client.getOutputStream(), true);
25.                     while (true) {//每一秒向客户端发送一个信息
26.                         ps.println((times++) + " Jack Mary");
27.                         Thread.sleep(1000);
28.                     }
29.                 } catch (Exception e) {
30.                     e.printStackTrace();
31.                 }
32.             }
33.         }
34. }
```

Spark Streaming 客户端如【代码 11-14】所示。

【代码 11-14】UpdateStateByKeyClient.scala

```
1. package org.hadoop.spark
2. object UpdateStateByKeyClient {
3.   def main(args: Array[String]): Unit = {
4.     val conf = new SparkConf()
5.       .setMaster("local[2]")
6.       .setAppName("NetworkWordCount")
7.     //声明 Spark Streaming 对象
8.     val ssc = new StreamingContext(conf, Seconds(1));
9.     //必须设置检查点
10.    ssc.checkpoint("file:///D:/a/cp");
11.    //声明 updateStateByKey 函数=(参数){函数体}
12.    /** 其中第一个参数，为前面累计的结果，第二个参数为新的结果 **/
13.    val updateStateByKey = (values: Seq[Int], state: Option[Int]) => {
14.      val newValue: Int = state.getOrElse(0);
15.      //如果没有则为 0
16.      //将前面的值进行累计求和
17.      val sum: Int = values.sum;
18.      Some(sum + newValue); //这是返回值
19.    };
20.    val rdd = ssc.socketTextStream("server201", 9999);
21.    //转换完成以后再执行 update 函数
```

```
22.      val rdd2 = rdd.flatMap(_.split("\\s+")).map((_, 1))
23.         .updateStateByKey[Int](updateStateByKey);
24.      rdd2.print();
25.      ssc.start();
26.      ssc.awaitTermination();
27.      Thread.sleep(1000 * 30);
28.      ssc.stop();
29.   }
30. }
```

先启动 Java 代码，再启动 Scala 代码，查看 Scala 方 Spark Stream 接收到的数据：

```
-------------------------------------------
Time: 1547973415000 ms    第一次连接，显示 Mary,Jack 都是 1 个
-------------------------------------------
(Mary,1)
(1,1)
(Jack,1)


-------------------------------------------
Time: 1547973416000 ms   第二个获取信息，显示 Mary,Jack 都是 2 个，后面以此类推
-------------------------------------------
(2,1)
(Mary,2)
(1,1)
(Jack,2)

-------------------------------------------
Time: 1547973417000 ms
-------------------------------------------
(2,1)
(3,1)
(Mary,3)
(1,1)
(Jack,3)
```

11.7 共享变量

共享变量是指在各个 executor 上都可以访问的变量。共享变量可以让所有 executor 更高效地访问程序共同使用的变量。

11.7.1 广播变量

广播变量（Broadcast Variable）被序列化以后，传递给每一个 executor，并缓存在 executor 里面，以后在需要访问时，高效地获取广播变量的数据。

通过 SparkContext 的 broadcast() 方法创建广播变量，它返回 BroadCast[T] 类型。

注意：对于广播变量 BroadCast[T] 类型，如果希望访问它的数据，则需要调用它的 value 属性。

同时，广播变量是单向传递的，即它只能从 driver 到 executor，所以，一个广播变量是没有办法更新的。如果希望更新，请使用累加器。

以下是没有使用广播变量的示例代码。

【代码 11-15】ShareVar.scala

```
1.  package org.hadoop.spark
2.  import org.apache.spark.{SparkConf, SparkContext}
3.  object ShareVar {
4.    def main(args: Array[String]): Unit = {
5.      val conf:SparkConf = new SparkConf();
6.      conf.setMaster("local[*]");
7.      conf.setAppName("ShareVariable");
8.      val sc:SparkContext = new SparkContext(conf);
9.      sc.setLogLevel("WARN");
10.     //声明变量，Map(key)其中 key 参数，是指根据 key 值查询 value 的数据
12.     val lookup = Map(1->"a",2->"b",3->"c",4->"d");
13.     val rdd = sc.parallelize(Seq(1,2,4));
14.     val rdd2 = rdd.map(lookup(_));
15.     println(rdd2.collect().toSet);
16.     sc.stop();
17.   }
18. }
```

现在将 lookup 变量声明为广播变量，可以让程序更高效的运行，如【代码 11-16】所示。

【代码 11-16】ShareVar2.scala

```
1.  package org.hadoop.spark
2.  import org.apache.spark.broadcast.Broadcast
3.  import org.apache.spark.{SparkConf, SparkContext}
4.  object ShareVar2 {
5.    def main(args: Array[String]): Unit = {
6.      val conf:SparkConf = new SparkConf();
7.      conf.setMaster("local[*]");
8.      conf.setAppName("ShareVariable");
9.      val sc:SparkContext = new SparkContext(conf);
10.     sc.setLogLevel("WARN");
11.     //声明广播变量
12.     val lookup:Broadcast[Map[Int,String]] =
13.       sc.broadcast[Map[Int,String]](Map(1->"a",2->"b",3->"c",4->"d"));
14.     val rdd = sc.parallelize(Seq(1,2,4));
15.     //访问访问变量的值，请使用.value
16.     val rdd2 = rdd.map(lookup.value(_));
17.     println(rdd2.collect().toSet);
18.     sc.stop();
19.   }
20. }
```

11.7.2 累加器

累加器（Accumulator）是任务中只能对它做加法操作的共享变量，类似于 MapReduce 中的计数器。当作业完成以后，driver 可以获取累加器的最终值。

通过 SparkContext 的 accumulator 来初始化一个累加器。

【代码 11-17】Accumulator.scala

```scala
1.  package org.hadoop.spark
2.  object Accumulator {
3.    def main(args: Array[String]): Unit = {
4.      val conf: SparkConf = new SparkConf();
5.      conf.setMaster("local[2]");
6.      conf.setAppName("Accumulator");
7.      val sc: SparkContext = new SparkContext(conf);
8.      sc.setLogLevel("WARN");
9.      //声明累加器
10.     val acc: LongAccumulator = sc.longAccumulator("A");
11.     val rdd = sc.parallelize(1 to 10);
12.     val rdd2 = rdd.map(i => {//注意 map 是惰性操作
13.       acc.add(1);//对累加器加 1
14.       i + 1;
15.     });
16.     println(rdd2.collect().toSet);
17.     println("输出结果:" + acc.count); //获取值=10
18.     sc.stop();
19.   }
20. }
```

11.8 本章小结

本章主要讲解了以下内容：

- Spark 本地运行和集群运行。
- Spark submit 提交 jar 并运行。
- Spark SQL。
- Spark Streaming、DStream、窗口函数等。
- 共享变量和累加变量。
- Scala 编程。

第 12 章

Spark 机器学习

本章主要内容：

- 机器学习概述。
- Spark 3.0 ML 介绍。
- 典型机器学习流程。
- 典型算法模型实战。

主流科技企业都在积极地将自己重定位于围绕 AI 和机器学习方向。机器学习是一门多领域交叉学科，涉及概率论、统计学、逼近论、凸分析、算法复杂度理论等多门学科。专门研究计算机怎样模拟或实现人类的学习行为，以获取新的知识或技能，重新组织已有的知识结构使之不断改善自身的性能。传统的机器学习模型是大部分 AI 应用背后的真正驱动力，而不是深度神经网络。工程师们仍然用着传统的软件和工具来实现机器学习工程，而后发现这些软件和工具并不是很有效：将数据传输到模型并最终输出结果的整个流水线最终只能产生一些零散的、互不兼容的东西。ML 是 Spark 提供的处理机器学习的功能库，该库包含许多机器学习算法，开发者可以不需要深入了解机器学习算法就能开发出相关程序。本章将介绍 Spark ML 基本知识以及使用方法。

12.1 机器学习

12.1.1 机器学习概述

机器学习是一类算法的总称，这些算法企图从大量历史数据中挖掘出其中隐含的规律，并用于预测或者分类。更具体地说，机器学习可以看作寻找一个函数，输入是样本数据，输出是期望的结果，只是这个函数过于复杂，以至于不太方便形式化表达。需要注意的是，机器学习的目标是使学到的函数很好地适用于"新样本"，而不仅仅是在训练样本上表现很好。学到的函数适用于新样本的能力，称为泛化（Generalization）能力。

机器学习的一般步骤：

步骤 01 选择一个合适的模型。这通常需要依据实际问题而定，针对不同的问题和任务需要选取恰当的模型，模型就是一组函数的集合。

步骤 02 判断一个函数的好坏。这需要确定一个衡量标准，也就是我们通常说的损失函数（Loss Function），损失函数的确定也需要依据具体问题而定，如回归问题一般采用欧氏距离，分类问题一般采用交叉熵代价函数。

步骤 03 找出"最好"的函数。如何从众多函数中最快地找出"最好"的那一个，这一步是最大的难点，做到又快又准往往不是一件容易的事情。常用的方法有梯度下降算法、最小二乘法等算法和其他一些技巧（tricks）。

学习得到"最好"的函数后，需要在新样本上进行测试，只有在新样本上表现很好，才算是一个"好"的函数。

基于其与经验、环境或者任何我们称之为输入数据的相互作用，一个算法可以用不同的方式对一个问题建模。流行的做法是首先考虑一个算法的学习方式。算法的主要学习方式和学习模型只有几个，我们将会逐一介绍它们，并且给出几个算法和它们适合解决的问题类型来作为例子。

- 监督学习：输入数据被称为训练数据，它们有已知的标签或者结果，比如垃圾邮件/非垃圾邮件或者某段时间的股票价格。模型的参数确定需要通过一个训练的过程，在这个过程中模型将会要求作出预测，当预测不符时，则需要作出修改。
- 无监督学习：输入数据不带标签或者没有一个已知的结果。通过推测输入数据中存在的结构来建立模型。这类问题的例子有关联规则学习和聚类。算法的例子比如 Apriori 算法和 K-means 算法。
- 半监督学习：输入数据由带标记的和不带标记的组成。合适的预测模型虽然已经存在，但是模型在预测的同时还必须有能力通过发现潜在的结构来组织数据。这类问题包括分类和回归。典型算法包括对一些其他灵活的模型的推广，这些模型都对如何给未标记数据建模作出了一些假设。
- 强化学习：输入数据作为来自环境的激励提供给模型，且模型必须作出反应。反馈并不像监督学习那样来自于训练的过程，而是作为环境的惩罚或者是奖赏。典型问题有系统和机器人控制。算法的例子包括 Q-学习和时序差分学习（Temporal Difference Learning）。当你处理大量数据来对商业决策建模时，通常会使用监督和无监督学习。

目前一个热门话题是半监督学习，比如会应用在图像分类中，涉及的数据集很大但是只包含极少数标记的数据。通常我们会把算法按照功能和形式的相似性来区分。比如树形结构和神经网络的方法。这是一种有用的分类方法，但也不是完美的。仍然有些算法很容易就可以被归入好几个类别，比如学习矢量量化，它既是受启发于神经网络的方法，又是基于实例的方法。也有一些算法的名字既描述了它处理的问题，也是某一类算法的名称，比如回归和聚类。正因为如此，你会从不同的来源看到对算法进行不同的归类。就像机器学习算法自身一样，没有完美的模型，只有足够好的模型。

随着大数据的发展，人们对大数据的处理要求也越来越高，原有的批处理框架 MapReduce 适合离线计算（在 MapReduce 中，由于其分布式特性——所有数据需要读写磁盘、启动 job 耗时较大，难以满足时效性要求），却无法满足实时性要求较高的业务，如实时推荐、用户行为分析等。

Spark 是一个类似于 MapReduce 的分布式计算框架，其核心是弹性分布式数据集，提供了比 MapReduce 更丰富的模型，可以快速地在内存中对数据集进行多次迭代，以支持复杂的数据挖掘算法和图形计算算法。

12.1.2 Spark ML

Spark 3.0 之后推荐使用 ML 库，而 MLlib 进入维护状态，不建议使用。ML 主要操作的是 DataFrame，而 MLlib 操作的是 RDD，相比于 MLlib 在 RDD 提供的基础操作，ML 在 DataFrame 上的抽象级别更高，数据和操作耦合度更低。

spark.ml 是 Spark 3.0 版本的机器学习库，提供了基于 DataFrame 高层次的 API，可以用来构建机器学习管道。之所以使用 spark.ml，是因为基于 DataFrame 的 API 更加的通用而且灵活。

有 4 种常见的机器学习问题：分类、回归、聚类和协同过滤，常见的算法类和工具类如图 12-1 所示。

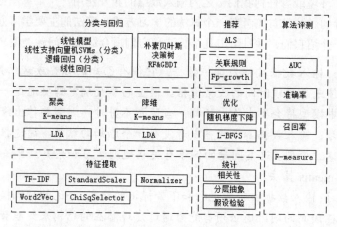

图 12-1

12.2 典型机器学习流程介绍

12.2.1 提出问题

图 12-2 是北京市海淀区的房价与面积关系数据，由这些数据估计出海淀区某个房子的售价大概是多少万元？可以从图中看到，用一些离散的点把刚才这些数据表示出来。图中横坐标是面积，纵坐标是售价。我们可以从图中看到一定的规律，如 200 平方米的面积售价在上千万元了。那机器学习就是用来做这样的事情：就是从已知的数据中找出规律，方便实现预测或者估计的功能。

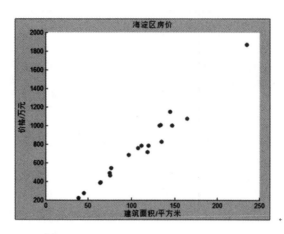

图 12-2

那么如何去找出规律？从图 12-3 可以看到，它上面的点基本都围绕在这个直线附近。从而可以看出 200 平方米的房价，基本上在 1400 万元到 1600 万元之间。所以机器学习就是要寻找这样的直线，从而找到房价的规律。

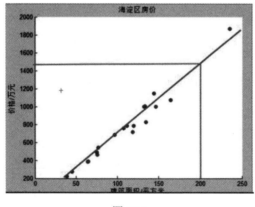

图 12-3

12.2.2 假设函数

通过上一小节我们初步判定样本点的分布趋向于一条直线的周边，我们可以考虑采用线性模型进行求解，首先就是要设定一个假设函数。平面几何里的这条直线可以用函数表示为：$h(x) = \theta_0 x_0 + \theta_1 x_1$。

如图 12-4 所示，横轴表示面积，纵轴表示房价，即公式为：房价=0.0+θ_1*面积，我们尝试绘制三条直线，其中 θ_0=0，θ_1 可以取值 6、7、8，$h(x)$ 最后算出来就是房价。这种函数我们把它叫作线性函数。

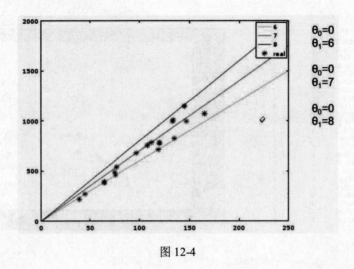

图 12-4

θ_0、θ_1 可以有更多的取值,它们取不同值时,与真实数据的吻合程度也有不同。比如最下面那条直线,可以看到位置和样本点之间还是有所偏差的,而 $\theta_1=7$ 时大部分房价是吻合的。那我们就可以以 $\theta_0=0$ 和 $\theta_1=7$ 来预测房价了。通过曲线观察就知道 $\theta_1=7$ 最好,但计算机怎么知道呢,答案就是通过计算损失函数获得。

12.2.3 代价函数

所谓代价函数,也叫损失函数,指的是用来判断预测函数 $h(x)$ 和真实数据之间的误差程度,如图 12-5 所示。

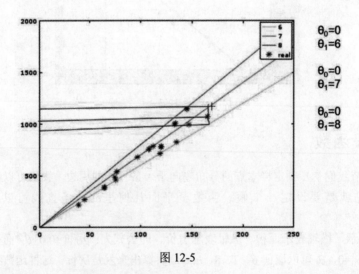

图 12-5

根据样本点的横坐标,代入到三条不同的直线函数 $h(x)$ 计算出房价,然后和真实房价进行比较。可以发现最下面那条直线的值更接近真实房价。

通过上面的分析,问题变成为寻找一个最接近真实规律的 θ_0 和 θ_1,最终我们就得到了下面的损失函数。损失函数有很多种表现方式,下面这个公式是一个比较好理解的方式:

$$J(\theta_0,\theta_1) = \frac{1}{2m}\sum(h(x_{(m,2)}) - y_{(m,1)})^2$$

- h 是预测函数。
- x 面积 feature（特征）θ_0 和 θ_1 两个特征。
- y 房价 lable（标签）真实的数据。
- （预测值-真实值）^{2} 求平均，误差比较小，放大这个误差，求平方。
- $\sum$ 求和函数把所有差值的平方求和最后求平均。
- m 就是真实样本数据的数。如果我们有 100 条数据，那么 m=100。

这个函数用来衡量预测值 $h(x_i)$ 与真实样本数据 y_i 之间的差距。

假设我们给定了 θ_0 和 θ_1，那这个结果值是大了好，还是小了好呢？当然是越小越好，越小说明预测值和真实值越接近。

12.2.4 训练模型确定参数

上一小节中给出了损失函数，接下来就是计算 θ_0 和 θ_1 了。我们可以针对每一组 θ_0 和 θ_1 来计算它的运算结果，不断改变 θ_0 和 θ_1 的值，观察哪个结果最小。有效的办法，就是不断尝试改变 θ_0 和 θ_1。如 θ_0 和 θ_1 都是 0 先试一遍计算的结果是多少，然后 θ_0 改成 1，再看刚才的结果谁更小。如果是小当然好，如果是大，就把这组结果排除，再试下一组。反正不断地尝试 θ_0 和 θ_1，让计算的结果最小。

上面这套规则以及样本数据就称为模型。

不断尝试改变 θ_0 和 θ_1 的值来求得最合适的 θ 的过程称为训练模型。

12.3 经典算法模型实战

机器学习算法主要包括分类、回归、聚类、推荐算法等典型算法模型，本节介绍一些常用经典模型的核心知识点及实战应用。

Spark 本地运行模式也被称为 Local[N]模式，是用单机的多个线程来模拟 Spark 分布式计算，直接运行在本地便于调试，通常用来验证开发出来的应用程序逻辑上有没有问题。其中 N 代表可以使用 N 个线程，每个线程拥有一个 core。如果不指定 N，则默认是 1 个线程（该线程有 1 个 core）。

本章涉及的 Spark 机器学习的代码均是在 Spark 本地模式上验证，这种模式方便读者进行快速开发和调试，详细源码参看本章的配套代码。

12.3.1 聚类算法实战

1. 算法知识点

聚类（Cluster analysis）有时也被翻译为簇类，其核心任务是：将一组目标 object 划分为若干个簇，每个簇之间的 object 尽可能相似，簇与簇之间的 object 尽可能相异。聚类算法是机器学习（或者说是数据挖掘更合适）中重要的一部分，除了最为简单的 K-means 聚类算法外，比较常见的还有层次法（CURE、CHAMELEON 等）、网格算法（STING、WaveCluster 等），等等。

较权威的聚类问题定义：所谓聚类问题，就是给定一个元素集合 D，其中每个元素具有 n 个可观察属性，使用某种算法将 D 划分成 k 个子集，要求每个子集内部的元素之间相异度尽可能低，而不同子集的元素相异度尽可能高。其中每个子集叫作一个簇。

K-means 聚类属于无监督学习，以往的回归、朴素贝叶斯、SVM 等都是有类别标签 y 的，也就是说，样例中已经给出了样例的分类。而聚类的样本中却没有给定 y，只有特征 x，比如假设宇宙中的星星可以表示成三维空间中的点集(x,y,z)。聚类的目的是找到每个样本 x 潜在的类别 y，并将同类别 y 的样本 x 放在一起。比如上面的星星例子，聚类后结果是一个个星团，星团里面的点相互距离比较近，星团间的星星距离就比较远了。

聚类与分类不同，分类是示例式学习，要求分类前明确各个类别，并断言每个元素映射到一个类别。聚类是观察式学习，在聚类前可以不知道类别甚至不给定类别数量，是无监督学习的一种。目前聚类广泛应用于统计学、生物学、数据库技术和市场营销等领域，相应的算法也非常多。

2. 案例说明

在本案例中我们将介绍 K-means 算法，K-means 属于基于平方误差的迭代重分配聚类算法，其核心思想十分简单：

（1）随机选择 K 个中心点。
（2）计算所有点到这 K 个中心点的距离，选择距离最近的中心点为其所在的簇。
（3）简单地采用算术平均数（mean）来重新计算 K 个簇的中心。
（4）重复步骤（2）和步骤（3），直至簇类不再发生变化或者达到最大迭代值。
（5）输出结果。

K-Means 算法的结果好坏依赖于对初始聚类中心的选择，容易陷入局部最优解，对 K 值的选择没有准则可依循，对异常数据较为敏感，只能处理数值属性的数据，聚类结构可能不平衡。

3. 数据处理及算法应用

本案例使用的数据为 sample_kmeans_data.txt，可以在配套代码项目根目录的 data/目录中找到。该文件提供了 6 个点的空间位置坐标，使用 K-means 聚类对这些点进行分类。

sample_kmeans_data.txt 的数据如下所示：

```
0.0 0.0 0.0
0.1 0.1 0.1
0.2 0.2 0.2
9.0 9.0 9.0
9.1 9.1 9.1
9.2 9.2 9.2
```

其中每一行都是一个坐标点的坐标值。

fit 方法是 ML 中 K-means 模型的训练方法，其内容如下：

```
class KMeans extends Estimator[KMeansModel] with KMeansParams with DefaultParamsWritable
    //Kmeans 类
    def fit(dataset: Dataset[_]): KMeansModel
    //训练的方法
```

若干个参数可由一系列 setter 函数来设置,参数解释如下:

- data: Dataset[_]:输入的数据集。
- setK(value: Int):聚类分成的数据集数。
- setMaxIter(value: Int):最大迭代次数。

聚类算法的应用如【代码 12-1】所示。

【代码 12-1】KMeansExample.scala

```
1.  import org.apache.spark.ml.clustering.KMeans
2.  import org.apache.spark.ml.evaluation.ClusteringEvaluator
3.  import org.apache.spark.sql.SparkSession
4.  object KMeansExample {
5.    def main(args: Array[String]): Unit = {
6.      val spark = SparkSession
7.        .builder                          //创建 Spark 会话
8.        .master("local")                  //设置本地模式
9.        .appName("K-means")               //设置名称
10.       .getOrCreate()                    //创建会话变量
11.     //读取数据
12.     val dataset = spark.read.format("libsvm").load("data/ sample_kmeans_data.txt")
13.     //训练模型,设置参数,载入训练集数据正式训练模型
14.     val kmeans = new KMeans().setK(3).setSeed(1L)
15.     val model = kmeans.fit(dataset)
16.     //使用测试集作预测
17.     val predictions = model.transform(dataset)
18.     //使用轮廓分评估模型
19.     val evaluator = new ClusteringEvaluator()
20.     val silhouette = evaluator.evaluate(predictions)
21.     println(s"Silhouette with squared euclidean distance = $silhouette")
22.     //展示结果
23.     println("Cluster Centers: ")
24.     model.clusterCenters.foreach(println)
25.     spark.stop()
26.   }
27. }
```

这个项目需要引入 Spark 核心包、机器学习包等依赖,本章项目对应的依赖配置具体如下。

【代码 12-2】pom.xml

```
1.  <dependencies>
2.      <dependency>
3.          <groupId>org.scala-lang</groupId>
4.          <artifactId>scala-library</artifactId>
5.          <version>2.12.7</version>
6.      </dependency>
7.      <dependency>
```

```
8.        <groupId>org.apache.spark</groupId>
9.        <artifactId>spark-core_2.12</artifactId>
10.        <version>3.1.1</version>
11.    </dependency>
12.    <!--引入 sparkStreaming 依赖-->
13.    <dependency>
14.        <groupId>org.apache.spark</groupId>
15.        <artifactId>spark-streaming_2.12</artifactId>
16.        <version>3.1.1</version>
17.    </dependency>
18.    <!--引入 sparkstreaming 整合 kafka 的依赖-->
19.    <dependency>
20.        <groupId>org.apache.spark</groupId>
21.        <artifactId>spark-streaming-kafka-0-8_2.11</artifactId>
22.        <version>2.0.2</version>
23.    </dependency>
24.    <dependency>
25.        <groupId>org.apache.spark</groupId>
26.        <artifactId>spark-mllib_2.12</artifactId>
27.        <version>3.1.1</version>
28.    </dependency>
29. </dependencies>
```

其中轮廓分数使用 ClusteringEvaluator，它测量一个簇中的每个点与相邻簇中点的接近程度，从而帮助判断簇是否紧凑且间隔良好。时间复杂度为 $O(tknm)$，其中 t 为迭代次数、k 为簇的数目、n 为样本点数、m 为样本点维度。空间复杂度为 $O(m(n+k))$，其中 k 为簇的数目、m 为样本点维度、n 为样本点数。K-means 是对三维数据进行聚类处理，如果是更高维的数据，请读者自行修改数据集进行计算和验证，程序的运行结果请自行打印验证。

12.3.2 回归算法实战

1. 算法知识点

线性回归是利用线性回归方程的函数，对一个或多个自变量和因变量之间关系进行建模的一种回归分析方法。只有一个自变量的情况称为简单回归，大于一个自变量情况的叫作多元回归，在实际情况中大多数都是多元回归。

线性回归（Linear Regression）问题属于监督学习（Supervised Learning）范畴，又称分类（Classification）或归纳学习（Inductive Learning）。这类分析中训练数据集给出的数据类型是确定的。机器学习的目标是，对于给定的一个训练数据集，通过不断地分析和学习产生一个联系属性集合和类标集合的分类函数（Classification Function）或预测函数（Prediction Function），这个函数称为分类模型（Classification Model）或预测模型（Prediction Model）。通过学习得到的模型可以是一个决策树、规格集、贝叶斯模型或一个超平面。通过这个模型可以对输入对象的特征向量进行预测或对对象的类标进行分类。

回归问题中通常使用最小二乘（Least Squares）法来迭代最优的特征中每个属性的比重，通过损失函数（Loss Function）或错误函数（Error Function）定义来设置收敛状态，即作为梯度下降算法的逼近参数因子。

2. 案例说明

线性回归分析的整个过程可以简单描述为如下三个步骤：

（1）寻找合适的预测函数，即上文中的 $h(x)$，用来预测输入数据的判断结果。这个过程是非常关键的，需要对数据有一定的了解或分析，知道或者猜测预测函数的"大概"形式，比如是线性函数还是非线性函数，若是非线性的，则无法用线性回归来得出高质量的结果。

（2）构造一个 Loss 函数（损失函数），该函数表示预测的输出（h）与训练数据标签之间的偏差，可以是二者之间的差（$h-y$）或者是其他的形式（如平方差开方）。综合考虑所有训练数据的"损失"，将 Loss 求和或者求平均值，记为 $J(\theta)$ 函数，表示所有训练数据预测值与实际类别的偏差。

（3）显然，$J(\theta)$ 函数的值越小表示预测函数越准确（即 h 函数越准确），所以这一步需要做的是找到 $J(\theta)$ 函数的最小值。寻找函数的最小值有不同的方法，Spark 中采用的是梯度下降法（stochastic gradient descent，SGD）。

3. 数据及算法应用

首先需要完成线性回归的数据准备工作。在 ML 中，线性回归的示例用来演示训练弹性网络（ElasticNet）、正则化线性回归模型、提取模型汇总统计信息，以及使用 ElasticNet 回归综合。它的学习目标是最小化指定的损失函数，并进行正则化。

线性回归是连续值预测，最终获得一个基于特征变量的连续函数作为预测模型，衡量模型的准确度一般采用均方根误差 RMSE。预测值和真实值的均方差越小，说明模型预测的效果越好。具体如【代码 12-3】所示。

【代码 12-3】LinearRegression.scala

```
import org.apache.spark.ml.regression.LinearRegression
import org.apache.spark.sql.SparkSession
object LinearRegressionWithElasticNetExample {
   def main(args: Array[String]): Unit = {
      val spark = SparkSession
        .builder                                              //创建 Spark 会话
        .master("local")                                      //设置本地模式
        .appName("LinearRegressionWithElasticNetExample")     //设置名称
        .getOrCreate()                                        //创建会话变量
      //$example on$
      //读取数据
      val training = spark.read.format("libsvm")
        .load("data/sample_linear_regression_data.txt")
      //建立一个 Estimator，并设置参数
      val lr = new LinearRegression()
        .setMaxIter(10)
        .setRegParam(0.3) //正则化参数
        .setElasticNetParam(0.8) //使用 ElasticNet 回归
      //训练模型
      val lrModel = lr.fit(training)
      //打印一些系数（回归系数表）和截距
```

```
        println(s"Coefficients: ${lrModel.coefficients} Intercept: ${lrModel.
intercept}")
        //汇总一些指标并打印结果和一些监控信息
        val trainingSummary = lrModel.summary
        println(s"numIterations: ${trainingSummary.totalIterations}")
        println(s"objectiveHistory: [${trainingSummary.objectiveHistory.mkStr
ing(",")}]")
        trainingSummary.residuals.show()
        println(s"RMSE: ${trainingSummary.rootMeanSquaredError}")
        println(s"r2: ${trainingSummary.r2}")
        spark.stop()
    }
}
```

其中，setElasticNetParam 设置的是 elasticNetParam，范围是 0~1（包括 0 和 1）。如果设置 0，则惩罚项是 L2 的惩罚项，训练的模型简化为 Ridge 回归模型；如果设置 1，那么惩罚项就是 L1 的惩罚项，等价于 Lasso 模型。

回归结果如下所示：

```
Coefficients: [0.0,0.32292516677405936,-0.3438548034562218,1.9156017023458414,0.05288058680386263,0.765967720459771,0.0,-0.15105392669186682,-0.21587930360904642,0.22025369188813426] Intercept: 0.1598936844239736
numIterations: 7
objectiveHistory: [0.4999999999999994,0.4967620357443381,0.4936361664340463,0.4936351537897608,0.4936351214177871,0.49363512062528014,0.4936351206216114]
+--------------------+
|           residuals|（残差）
+--------------------+
|  -9.889232683103197|
|  0.5533794340053554|
|  -5.204019455758823|
| -20.566686715507508|
|   -9.4497405180564|
|  -6.909112502719486|
|  -10.00431602969873|
|   2.062397807050484|
|   3.1117508432954772|
| -15.893608229419382|
|  -5.036284254673026|
|   6.483215876994333|
|  12.429497299109002|
|  -20.32003219007654|
|  -2.0049838218725005|
| -17.867901734183793|
|   7.646455887420495|
|  -2.2653482182417406|
| -0.10308920436195645|
|  -1.380034070385301|
+--------------------+
only showing top 20 rows
```

```
RMSE: 10.1890771675984750
r2: 0.0228614669139581840
```

上述结果中，r2 表示的是判定系数，也称为拟合优度，越接近 1 越好。

12.3.3 协同过滤算法实战

1. 算法说明

协同过滤（Collaborative Filtering，CF），WIKI 上的定义是：简单来说是利用某个兴趣相投、拥有共同经验之群体的喜好来推荐感兴趣的信息给使用者，个人通过合作的机制给予信息相当程度的回应（如评分）并记录下来以达到过滤的目的，进而帮助别人筛选信息，回应不一定局限于特别感兴趣的，特别不感兴趣信息的记录也相当重要。

协同过滤常被应用于推荐系统。这些技术旨在补充"用户—商品"关联矩阵中所缺失的部分。

基于模型的协同过滤，用户和商品通过一小组隐性因子进行表达，并且这些因子也用于预测缺失的元素。ML 使用交替最小二乘法（ALS）来学习这些隐性因子。

这里对最小二乘法 ALS 做一下说明。ALS 是 Alternating Least Squares 的缩写，意为交替最小二乘法；而 ALS-WR 是 Alternating Least Squares with Weighted-λ-Regularization 的缩写，意为加权正则化交替最小二乘法。该方法常用于基于矩阵分解的推荐系统中，比如用户对商品的评分矩阵，可以分解为一个用户对隐含特征偏好的矩阵，一个是商品所包含的隐含特征的矩阵；对于 $R(m \times n)$ 的矩阵，ALS 旨在找到两个低维矩阵 $X(m \times k)$ 和矩阵 $Y(n \times k)$，来近似逼近 $R(m \times n)$，在这过程中把用户评分缺失项填上，并根据这个分数给用户推荐。

用户对物品或者信息的偏好，根据应用本身的不同，可能包括用户对物品的评分、用户查看物品的记录、用户的购买记录等。其实这些用户的偏好信息可以分为两类：

- 显式的用户反馈：这类是用户在网站上自然浏览或者使用网站以外，显式地提供反馈信息，例如用户对物品的评分或者对物品的评论。
- 隐式的用户反馈：这类是用户在使用网站时产生的数据，隐式地反映了用户对物品的喜好，例如用户购买了某物品，用户查看了某物品的信息，等等。

显式的用户反馈能准确地反映用户对物品的真实喜好，但需要用户付出额外的代价；而隐式的用户行为，通过一些分析和处理，也能反映用户的喜好，只是数据不是很精确，有些行为的分析存在较大的噪音。但只要选择正确的行为特征，隐式的用户反馈也能得到很好的效果，只是行为特征的选择可能在不同的应用中有很大的不同，例如在电子商务的网站上，购买行为其实就是一个能很好表现用户喜好的隐式反馈。

推荐引擎根据不同的推荐机制可能用到数据源中的一部分，然后根据这些数据，分析出一定的规则，或者直接对用户对其他物品的喜好进行预测计算。这样推荐引擎可以在用户进入时给他推荐他可能感兴趣的物品。

Spark ML 目前支持基于协同过滤的模型，在这个模型里，用户和产品被一组可以用来预测缺失项目的潜在因子来描述。特别是我们实现交替最小二乘（ALS）算法来学习这些潜在的因子。在 ML 中的实现有如下参数：

- numBlocks：是用于并行化计算的分块个数（设置为-1 时，为自动配置）。
- rank：是模型中隐性因子的个数。
- iterations：是迭代的次数。
- lambda：是 ALS 的正则化参数。
- implicitPrefs：决定了是用显性反馈 ALS 的版本，还是用隐性反馈数据集的版本。
- alpha：是一个针对隐性反馈 ALS 版本的参数，这个参数决定了偏好行为强度的基准。

算法原理图如图 12-6 所示。

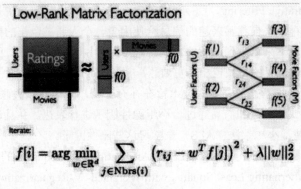

图 12-6

2. 案例说明

我们来看 Spark 3.0 ML 中 ALS 算法的程序设计。

（1）切分数据集

ALS 算法的前验基础是切分数据集，首先建立数据集文件 sample_movielens_ratings.txt，内容如图 12-7 所示。

```
sample_movielens_ratings.txt - 记事本
文件(F) 编辑(E) 格式(O) 查看(V) 帮助(H)
0::2::3::1424380312
0::3::1::1424380312
0::5::2::1424380312
0::9::4::1424380312
0::11::1::1424380312
0::12::2::1424380312
0::15::1::1424380312
0::17::1::1424380312
0::19::1::1424380312
0::21::1::1424380312
0::23::1::1424380312
0::26::3::1424380312
0::27::1::1424380312
0::28::1::1424380312
0::29::1::1424380312
0::30::1::1424380312
0::31::1::1424380312
0::34::1::1424380312
0::37::1::1424380312
0::41::2::1424380312
0::44::1::1424380312
0::45::2::1424380312
```

图 12-7

需要注意的是，ML 中的 ALS 算法有固定的数据格式，源码如下：

```
case class Rating(userId: Int, movieId: Int, rating: Float, timestamp: Long)
```

其中，Rating 是固定的 ALS 输入格式，要求是一个元组类型的数据，其中的数值分别为 [Int,Int, Float, Long]，因此在数据集建立时用户名和物品名分别用数值代替，而最终的评分没有变化，最后是一个时间戳（类型是 Long）。基于 Spark 3.0 架构，我们可以将迭代算法 ALS 很好地并行化处理。

（2）建立 ALS 数据模型

第二步就是建立 ALS 数据模型。ALS 数据模型是根据数据集训练获得的，源码如下：

```
val Array(training, test) = ratings.randomSplit(Array(0.8, 0.2))
val als = new ALS()
  .setMaxIter(5)
  .setRegParam(0.01)
  .setUserCol("userId")
  .setItemCol("movieId")
  .setRatingCol("rating")
val model = als.fit(training)
```

ALS 是由若干个 setter 设置参数构成的，其解释如下：

- numBlocks（numItemBlocks、numUserBlocks）：并行计算的 block 数（-1 为自动配置）。
- rank：模型中的隐藏因子数。
- maxIter：算法最大迭代次数。
- regParam：ALS 中的正则化参数。
- implicitPref：使用显示反馈 ALS 变量或隐式反馈。
- alpha：ALS 隐式反馈变化率用于控制每次拟合修正的幅度。
- coldStartStrategy：将 coldStartStrategy 参数设置为 "drop"，以便删除 DataFrame 包含 NaN 值的预测中的任何行。

这些参数协同作用，从而控制 ALS 算法的模型训练。

最终 Spark 3.0 ML 库基于 ALS 算法的协同过滤推荐代码如【代码 12-4】所示。

【代码 12-4】MovieLensALS.scala

```
import org.apache.spark.ml.evaluation.RegressionEvaluator
import org.apache.spark.ml.recommendation.ALS
import org.apache.spark.sql.SparkSession
object ALSExample {
  //定义 Rating 格式
  case class Rating(userId: Int, movieId: Int, rating: Float, timestamp: Long)
  def parseRating(str: String): Rating = {
    val fields = str.split("::")//分隔符
    assert(fields.size == 4)
    Rating(fields(0).toInt,   fields(1).toInt,   fields(2).toFloat,
fields(3).toLong)
```

```scala
        }
      def main(args: Array[String]): Unit = {
        val spark = SparkSession
          .builder                            //创建 Spark 会话
          .master("local")                    //设置本地模式
          .appName("ALSExample")              //设置名称
          .getOrCreate()                      //创建会话变量
        import spark.implicits._
        //读取 Rating 格式并转换 DF
        val ratings = spark.read.textFile("data/als/sample_movielens_ratings.txt")
          .map(parseRating)
          .toDF()
        val Array(training, test) = ratings.randomSplit(Array(0.8, 0.2))
        //在训练集上构建推荐系统模型、ALS 算法,并设置各种参数
        val als = new ALS()
          .setMaxIter(5)
          .setRegParam(0.01)
          .setUserCol("userId")
          .setItemCol("movieId")
          .setRatingCol("rating")
        val model = als.fit(training)//得到一个 model,一个 Transformer
        //在测试集上评估模型,标准为 RMSE
        //设置冷启动的策略为'drop',以保证不会得到一个'NAN'的预测结果
        model.setColdStartStrategy("drop")
        val predictions = model.transform(test)
        val evaluator = new RegressionEvaluator()
          .setMetricName("rmse")
          .setLabelCol("rating")
          .setPredictionCol("prediction")
        val rmse = evaluator.evaluate(predictions)
        println(s"Root-mean-square error = $rmse")
        //为每一个用户推荐10部电影
        val userRecs = model.recommendForAllUsers(10)
        //为每部电影推荐10个用户
        val movieRecs = model.recommendForAllItems(10)
        //为指定的一组用户生成 top10 个电影推荐
        val users = ratings.select(als.getUserCol).distinct().limit(3)
        val userSubsetRecs = model.recommendForUserSubset(users, 10)
        //为指定的一组电影生成 top10 个用户推荐
        val movies = ratings.select(als.getItemCol).distinct().limit(3)
        val movieSubSetRecs = model.recommendForItemSubset(movies, 10)
        //打印结果
        userRecs.show()
        movieRecs.show()
        userSubsetRecs.show()
        movieSubSetRecs.show()
        spark.stop()
      }
    }
```

在上面的程序中,使用 ALS()根据已有的数据集建立了一个协同过滤矩阵推荐模型,之后使用 recommendForAllUsers 方法为一个用户推荐 10 个物品(电影)等,结果打印如下:

```
Root-mean-square error = 1.684832316936912
```

```
+------+--------------------+
|userId|     recommendations|
+------+--------------------+
|    28|[[25, 6.00149], [...|
|    26|[[94, 5.29422], [...|
|    27|[[47, 6.3299623],...|
|    12|[[46, 6.5864477],...|
|    22|[[7, 5.437798], [...|
|     1|[[68, 3.8732295],...|
|    13|[[96, 3.8646204],...|
|     6|[[25, 4.5257554],...|
|    16|[[85, 4.960823], ...|
|     3|[[96, 4.1602864],...|
|    20|[[22, 4.770223], ...|
|     5|[[55, 4.090011], ...|
|    19|[[46, 5.232961], ...|
|    15|[[46, 4.8397903],...|
|    17|[[90, 4.914645], ...|
|     9|[[48, 5.1486597],...|
|     4|[[52, 4.2062426],...|
|     8|[[29, 5.071128], ...|
|    23|[[90, 5.842731], ...|
|     7|[[27, 5.47984], [...|
+------+--------------------+
only showing top 20 rows

+-------+--------------------+
|movieId|     recommendations|
+-------+--------------------+
|     31|[[12, 3.931116], ...|
|     85|[[16, 4.960823], ...|
|     65|[[23, 4.9316106],...|
|     53|[[21, 5.080318], ...|
|     78|[[0, 1.5588677], ...|
|     34|[[18, 4.6249347],...|
|     81|[[28, 4.7397876],...|
|     28|[[24, 5.2909055],...|
|     76|[[0, 4.9046974], ...|
|     26|[[11, 4.4119563],...|
|     27|[[7, 5.47984], [1...|
|     44|[[24, 4.7212014],...|
|     12|[[28, 4.688432], ...|
|     91|[[11, 3.1263103],...|
|     22|[[26, 5.134186], ...|
|     93|[[2, 5.194844], [...|
|     47|[[27, 6.3299623],...|
```

```
|     1|[[25, 2.9610748],...|
|    52|[[14, 4.997468], ...|
|    13|[[23, 3.8639143],..|
+------+--------------------+
only showing top 20 rows

+------+--------------------+
|userId|     recommendations|
+------+--------------------+
|    28|[[25, 6.00149], [...|
|    26|[[94, 5.29422], [...|
|    27|[[47, 6.3299623],...|
+------+--------------------+

+-------+--------------------+
|movieId|     recommendations|
+-------+--------------------+
|     31|[[12, 3.931116], ...|
|     85|[[16, 4.960823], ...|
|     65|[[23, 4.9316106],...|
+-------+--------------------+
```

在使用 ALS 进行预测时，通常会遇到测试数据集中的用户或物品没有出现的情况，这些用户或物品在训练模型期间不存在。针对上述问题，Spark 提供了将 coldStartStrategy 参数设置为"drop"的方式，就是删除 DataFrame 中包含 NaN 值的预测中的任何行。然后根据非 NaN 数据对模型进行评估，并且该评估是有效的。目前支持的冷启动策略是"nan"（上面提到的默认值）和"drop"，将来可能会支持进一步的策略。

提示：程序中的 rank 表示隐藏因子，numIterator 表示循环迭代的次数，读者可以根据需要调节数值。报出 StackOverFlow 错误时，可以适当地调节虚拟机或者 IDE 的栈内存。另外，读者可以尝试调用 ALS 中的其他方法，以更好地理解 ALS 模型的用法。Spark 官方实现的 ALS 由于调度方面的问题在训练的时候比较慢。

第 13 章

影评分析项目实战

本章主要内容：

- 网络爬虫的基本架构。
- 自然语言处理的入门级应用。
- 基于 Hadoop 的数据清洗。
- 使用 MapReduce 进行数据处理。
- 使用词云工具生成数据可视化的图像。

本章将通过一个综合案例即影评分析实战项目，介绍大数据分析处理的一般流程。通过项目流程的详细实现，掌握数据分析每个流程节点的关键技术及应用，包括编写爬虫程序进行数据采集，采用分词技术进行评论数据分析，采用 MapReduce 进行数据处理，以及进行数据结果的可视化呈现。为了方便调试项目，本章采用 Hadoop 本地运行模式，项目代码在读者本地计算机即可运行。

13.1 项目内容

本项目主要包含大数据环境中数据爬取的应用、自然语言处理的应用以及 MapReduce 的应用，以实现电影影评分析程序，帮助读者了解大数据体系架构的开发流程，以及利用现有技术解决生活中遇到的问题。本项目大致分析流程为：通过 Jsoup 爬取电影影评数据，将爬取的数据通过 Ikanalyzer 进行自然语言分词处理，将分词的数据通过 MapReduce 进行数据清洗达到 kumo 做词云图形化展示所要求的数据格式，从而实现从数据爬取→数据分析→数据可视化的完整流程。

13.2　项目需求及分析

（1）通过 Jsoup 爬取电影影评网站的数据，Jsoup 是一个开源 Java 库，用于解析、提取和操作存储在 HTML 文档中的数据，既然是解析、提取 HTML 中的数据，首先要了解要爬取网站页面的内容，在浏览器输入 https://movie.douban.com/ 浏览豆瓣电影页面，如图 13-1 所示。

图 13-1

选择想要分析的电影，本实验以电影《惊奇队长》为例，点击进入该电影的简介页面，将页面向下拖动，会看到"惊奇队长短评"一栏，如图 13-2 所示。

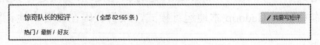

图 13-2

点击"全部 xxx 条"链接，跳转到该电影的影评页面，如图 13-3 所示。

图 13-3

将影评页面往下拖曳到最后，看到"后页"按钮，点击后查看地址栏，如图 13-4 所示。

图 13-4

- 26213252：表示电影的 id。
- start=20：表示从第几个影评开始。
- limit=20：表示显示多少个影评。

点击"前页"按钮后观察地址栏变化，如图 13-5 所示。

图 13-5

通过观察地址栏的变化得出结论，改变 start 的值控制地址栏的变化来获取不同页面的影评数据。在电影短评页面按 F12 键，进入开发者工具页面查看页面的 HTML 代码，如图 13-6 所示。

图 13-6

按 Ctrl+Shift+C 组合键后，开发者工具页面左上角的一个小图标会变成蓝色，如图 13-7 所示。

图 13-7

将鼠标指针拖动到左侧的原页面位置，点击任意一条评论，在右侧的开发者工具页面会自动跳转到该条评论对应的 HTML 代码区域，如图 13-8 所示。

图 13-8

通过图 13-8 发现 HTML 代码中存储评论数据的是 span 标签，class 为 short。点击标签前的小三角，展开标签可查看 HTML 中的评论数据，如图 13-9 所示。

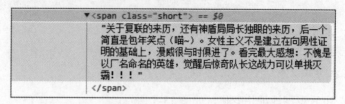

图 13-9

通过上述分析可得出结论：获取每页 HTML 中 class 为 short 的 span 标签，即可获取每页的评论数据。将这些数据写入到一个文本文件中，如图 13-10 所示。

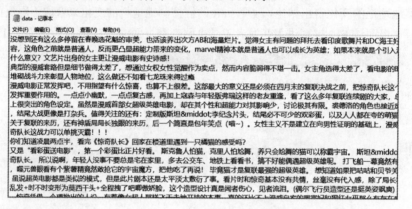

图 13-10

（2）在上一步爬取完数据的基础上，对影评数据文本利用分词器进行自然语言处理。本实验用到 Ikanalyzer 工具。

IK Analyzer 是一个开源的、基于 Java 语言开发的轻量级的中文分词工具包。从 2006 年 12 月推出 1.0 版开始，IKAnalyzer 已经推出了 4 个大版本。最初，它是以开源项目 Luence 为应用主体，结合词典分词和文法分析算法的中文分词组件。从 3.0 版本开始，IK 发展为面向 Java 的公用分词组件，独立于 Lucene 的项目，同时提供了对 Lucene 的默认优化实现。在 2012 版本中，IK 实现了简单的分词歧义排除算法，标志着 IK 分词器从单纯的词典分词向模拟语义分词衍化。

将爬取的影评数据进行分词处理，提取特征词语，对分词后的数据进一步分析。例如：

文本原文：

IK-Analyzer 是一个开源的，基于 Java 语言开发的轻量级的中文分词工具包。从 2006 年 12 月推出 1.0 版开始，IKAnalyzer 已经推出了 3 个大版本。

分词结果：

ik-analyzer｜是｜一个｜一｜个｜开源｜的｜基于｜Java｜语言｜开发｜的｜轻量级｜量级｜的｜中文｜分词｜工具包｜工具｜从｜2006｜年｜12｜月｜推出｜1.0｜版｜开始｜ikanalyzer｜已经｜推出｜出了｜3｜个大｜个｜版本

停用词：在分词的结果中可能会出现大量无用词汇，例如上边的分词结果中出现了"是"这个词，如果在分词结果中不想看到这样的分词，我们可以在分词程序中添加停用词表，那么程序就会在分词时自动过滤这些在停用词表中提到的词汇。本项目我们使用哈工大提供的停用词表。

拓展词：如果文章中出现了生僻词汇，那么默认的分词程序不会将这些词汇单独分出来，我们需要在程序中添加一个拓展词表，例如上边提到的例子，我们想将"是一个"划分为一个词汇，那么在拓展词表中就要加入这个词汇，便于程序去读取。

最终将分词后的数据输出到结果文件中，如图 13-11 所示。

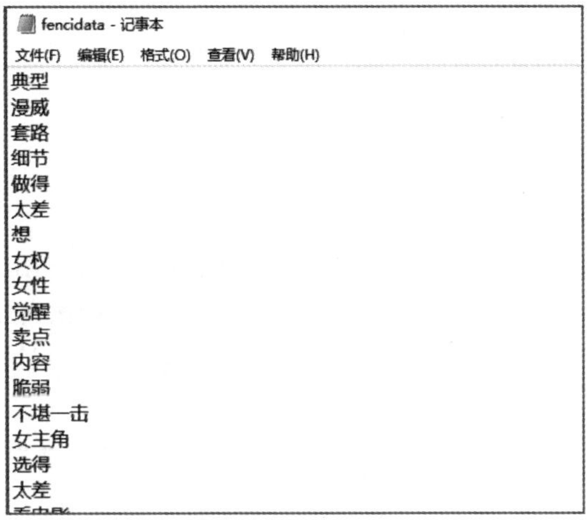

图 13-11

（3）接下来，通过 MapReduce 程序对分词后的结果进行统计并格式化输出。这里，我们需要在 MapReduce 程序中设置两个 job，将默认 key value 输出格式设置为 key:value 的输出格式，其中一个 job 用来完成词汇的统计，统计分词结果中每个词汇出现的次数，如图 13-12 所示。

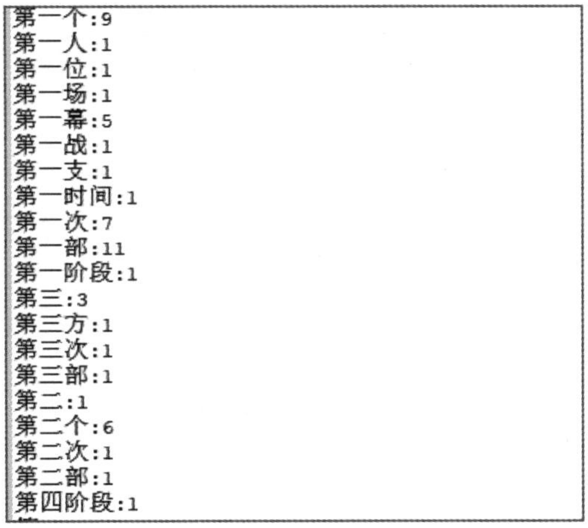

图 13-12

此时的数据是按照 MapReduce 默认排序输出的结果，后续词云展示用到的数据格式为"1:第一战"，此时的输出结果并不符合我们的需求，所以对第一次 job 输出的结果文件进行第二次 job 处理，将第一次输出的 key 和 value 调换位置，并且将出现次数多的词汇排在最前边，对 key 进行逆序排序。

为了方便最后词云展示的效果，将数据中单个词汇进行过滤，第二次 job 的输出结果如图 13-13 所示。

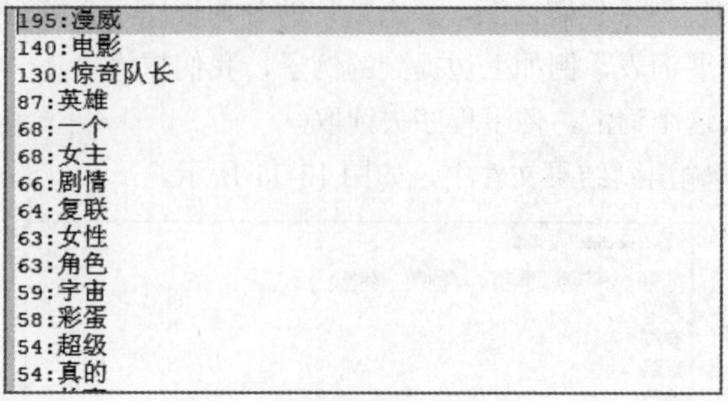

图 13-13

（4）最终将 MapReduce 的输出结果通过 Kumo 程序生成词云。Kumo 的目标是用 Java 创建功能强大且用户友好的 Word Cloud API。Kumo 可以直接生成一个图像文件。Word Cloud 是关键词的视觉化描述，用于汇总用户生成的标签或一个网站的文字内容。最终输出关于电影《惊奇队长》的词云图像如图 13-14 所示。

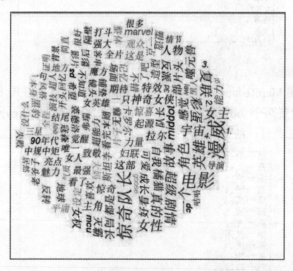

图 13-14

13.3 详细实现

13.3.1 搭建项目环境

打开 Eclipse 选择 File→New→Other→Maven Project 创建 Maven 工程，如图 13-15 所示。单击 Next 按钮，会进入"New Maven Project"界面，如图 13-16 所示。

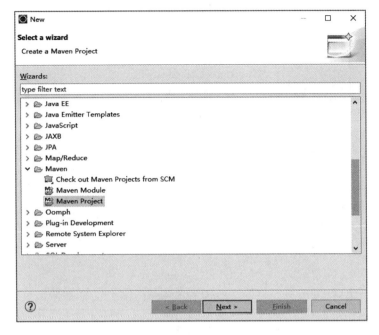

图 13-15

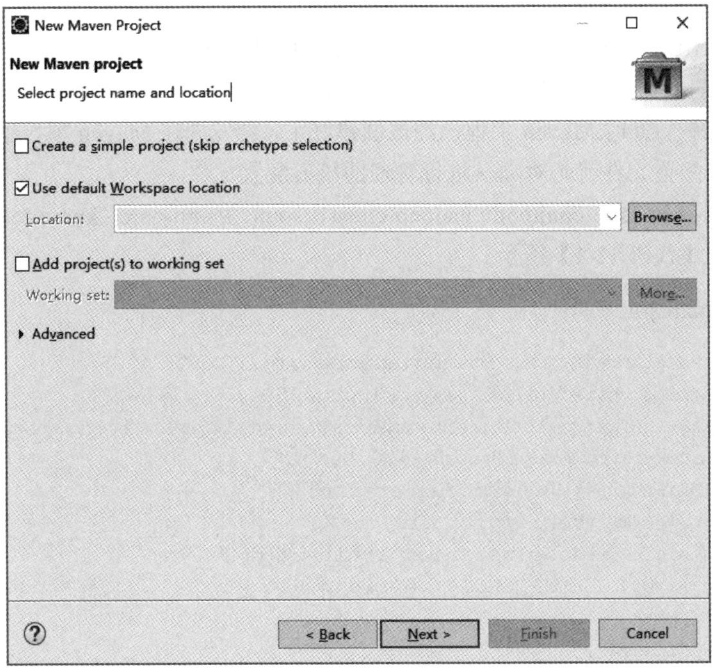

图 13-16

在图 13-16 中,"Create a simple project(skip archetype selection)"表示创建一个简单的项目(跳过对原型模板的选择),此处勾选"User default Workspace location"表示使用本地默认的工作空间。之后,单击 Next 按钮,结果如图 13-17 所示。

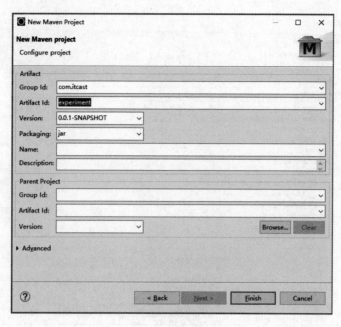

图 13-17

在图 13-17 中，GroupID 也就是项目组织唯一的标识符，实际对应 Java 的包结构，这里输入 com.mrchi。ArtifactID 就是项目的唯一标识符，实际对应项目的名称就是项目根目录的名称，这里输入 experiment，打包方式这里选择 jar 包方式即可，后续创建 Web 工程选择 war 包。

单击 Finish 按钮，此时 Maven 工程已经被创建好了，会发现在 Maven 项目中，有一个 pom.xml 的配置文件，这个配置文件就是对项目进行管理的核心配置文件。

本项目需要用到 hadoop-common、hadoop-client、jsoup、kumo-core、kumo-tokenizers、ikanalyzer 六种依赖，具体如【代码 13-1】所示。

【代码 13-1】pom.xml

```
1.   <project xmlns="http://maven.apache.org/POM/
2.   4.0.0" xmlns:xsi="http://www.w3.org/2001/
3.   XMLSchema-instance" xsi:schemaLocation="http://maven.apache.org/POM/4.0.0 http://maven.apache.org/xsd/maven-4.0.0.xsd">
4.       <modelVersion>4.0.0</modelVersion>
5.       <groupId>com.mrchi</groupId>
6.       <artifactId>hadoopInstance</artifactId>
7.       <version>0.0.1-SNAPSHOT</version>
8.
9.   <dependencies>
10.      <dependency>
11.          <groupId>org.jsoup</groupId>
12.          <artifactId>jsoup</artifactId>
13.          <version>1.7.3</version>
14.      </dependency>
15.      <dependency>
16.          <groupId>com.kennycason</groupId>
17.          <artifactId>kumo-core</artifactId>
```

```xml
18.         <version>1.17</version>
19.     </dependency>
20.     <dependency>
21.         <groupId>com.kennycason</groupId>
22.         <artifactId>kumo-tokenizers</artifactId>
23.         <version>1.17</version>
24.     </dependency>
25.     <dependency>
26.         <groupId>com.github.magese</groupId>
27.         <artifactId>ik-analyzer</artifactId>
28.         <version>7.7.1</version>
29.     </dependency>
30.     <dependency>
31.         <groupId>org.apache.hadoop</groupId>
32.         <artifactId>hadoop-common</artifactId>
33.         <version>3.2.2</version>
34.     </dependency>
35.     <dependency>
36.         <groupId>org.apache.hadoop</groupId>
37.         <artifactId>hadoop-client</artifactId>
38.         <version>3.2.2</version>
39.     </dependency>
40. </dependencies>
41. </project>
```

当添加依赖完毕后，Hadoop 相关 Jar 包就会自动下载，部分 Jar 包如图 13-18 所示。

```
> Maven: org.apache.hadoop:hadoop-annotations:3.2.2
> Maven: org.apache.hadoop:hadoop-auth:3.2.2
> Maven: org.apache.hadoop:hadoop-client:3.2.2
> Maven: org.apache.hadoop:hadoop-common:3.2.2
> Maven: org.apache.hadoop:hadoop-hdfs-client:3.2.2
> Maven: org.apache.hadoop:hadoop-mapreduce-client-common:3.2.2
> Maven: org.apache.hadoop:hadoop-mapreduce-client-core:3.2.2
> Maven: org.apache.hadoop:hadoop-mapreduce-client-jobclient:3.2.2
> Maven: org.apache.hadoop:hadoop-yarn-api:3.2.2
> Maven: org.apache.hadoop:hadoop-yarn-client:3.2.2
> Maven: org.apache.hadoop:hadoop-yarn-common:3.2.2
```

图 13-18

13.3.2 编写爬虫类

首先在项目 src 文件夹下创建 com.mrchi.move 包，在该路径下编写爬虫类 moveJsoup，如【代码 13-2】所示。

【代码 13-2】moveJsoup.java

```
1.  public class moveJsoup {
2.      public static void main(String[] args) throws InterruptedException, IOException {
```

```
3.      int num = 0;
4.      //爬取数据存储到本地的地址
5.      File fileNmame = new File("D:\\MoveData\\data.txt");
6.      BufferedWriter out = new BufferedWriter(new FileWriter(fileNmame));
7.      //爬取 25 页的影评数据,每页 20 条数据
8.      for (int i = 0; i < 25; i++) {
9.          String strNum = Integer.toString(num);
10.         //爬取数据的 url 地址
11.         String url =
12.             "https://movie.douban.com/subject/26213252/comments?start="+ num +"&amp;limit=20&amp;sort=new_score&amp;status=P";
13.         Connection connect = Jsoup.connect(url);
14.         Document document =
15.             connect.userAgent("浏览器的 User-Agentt")
16.                 .cookie("Cookie", "自己的 cookie")
17.                 //超时请求的时间
18.                 .timeout(6000)
19.                 //解析响应时忽略文档的 Content-Type
20.                 .ignoreContentType(true)
21.                 //以 GET 身份执行请求,并解析结果
22.                 .get();
23.         //获取 span 标签中 calss 等于 short 的内容
24.         Elements elements = document.select("span[class=short]");
25.         for (Element e : elements) {
26.             //将页面中的内容循环读取写入到文件 data.txt 中
27.             out.write(e.toString().replaceAll("</?[^>]+>", "")+"\r\n");
28.         }
29.         num = Integer.parseInt(strNum);
30.         num+=20;
31.     }
32.     //将缓冲区的数据输出
33.     out.flush();
34.     //关闭连接
35.     out.close();
36.  }
37. }
```

获取 User-Agent 和 Cookie 的方法:使用自己的账号密码登录豆瓣网后,按照项目需求中提到的步骤进入到全部短评页面,按 F12 键进入到开发者工具模式,然后单击地址栏中的 Network,如图 13-19 所示。

图 13-19

按 F5 键刷新页面，在 Network 下会出现一个文件，单击该文件会显示详细内容，如图 13-20 所示。

将文件内容拖到最底部，可以看到 Cookie 和 User-Agent 的内容如图 13-21 所示，将 ":" 号后的内容粘贴到代码中对应的位置即可。

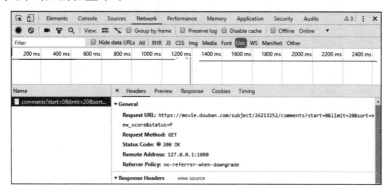

图 13-20

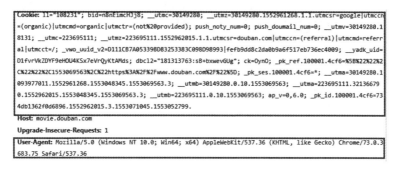

图 13-21

注意：Cookie 复制到代码中时会报错，因为 Cookie 中包含 """，处理的方法是在双引号的前边加上\转义符即可，例如\"asdasd\"。

如果不想使用用户名密码登录豆瓣网，那么可将代码中 25 换成 10，因为不登录账号的话只能查看 200 条评论数据。此时，将代码中.Cookie()一行删除即可。User-Agent 在不登录的情况下也可以用相同的方式获取。

第 38 行的正则去除字符串中的类似于<em></em>的标签，\r\n 用于每一条数据间做换行处理。

13.3.3　编写分词类

在 com.mrchi.move 包下编写分词类 moveFenci，如【代码 13-3】所示。

【代码 13-3】moveFenci.java

```
1.  public class moveFenci {
2.      public static void main(String[] args) throws IOException {
3.          //创建分词器对象
4.          Analyzer analyzer = null;
5.          //创建 TokenStream 流对象，分词的所有信息都会从 TokenStream 流中获取
```

```
6.      TokenStream ts = null;
7.      //创建 InputStreamReader 对象,将字节流转换为字符流
8.      InputStreamReader reader = null;
9.      //创建 BufferedWriter 对象,将文本写入字符输出流
10.     BufferedWriter writer = null;
11.     //创建 BufferedReader 对象,从字符输入流中读取文本
12.     BufferedReader br = null;
13.
14.     try{
15.         //分词结果输出的路径
16.         File fenCidata = new File("D:\\MoveData\\fencidata.txt");
17.         //影评信息输入的路径
18.         File moveData = new File("D:\\MoveData\\data.txt");
19.         //读取文件的数据,将字节流向字符流的转换
20.         reader = new InputStreamReader(new FileInputStream(moveData));
21.         //从字符流中读取文本
22.         br = new BufferedReader(reader);
23.         //创建了一个字符写入流的缓冲区对象,并和指定要被缓冲的流对象相关联
24.         writer = new BufferedWriter(new FileWriter(fenCidata));
25.         String line = "";
26.         //读取一行数据
27.         line = br.readLine();
28.         //判断文本是否有数据
29.         while (line != null) {
30.             line = br.readLine();
31.             if(line == null){
32.                 continue;
33.             }
34.             //构造分词对象,true 表示使用智能分词
35.             analyzer = new IKAnalyzer(true);
36.             //""指文档的域名,一片文档包含多个域名,这里任意指定即可
37.             ts = analyzer.tokenStream("", line);
38.             //获取每个单词信息
39.         CharTermAttribute term=ts.getAttribute(CharTermAttribute.class);
40.             //TokenStream 的流程需包含 reset()重置
41.             ts.reset();
42.             //遍历分词数据
43.             while(ts.incrementToken()){
44.                 //将数据写入到缓冲区中
45.                 writer.write(term.toString()+"\r\n");
46.             }
47.
48.         }
49.     }
50.     catch(Exception e){
51.         e.printStackTrace();
52.     }finally{
53.         //关闭流
```

```
54.            analyzer.close();
55.            //关闭流
56.            reader.close();
57.            //关闭流
58.            br.close();
59.            //刷新该流中的缓冲。将缓冲数据写到目的文件中去
60.            writer.flush();
61.            //关闭此流，再关闭前会先刷新
62.            writer.close();
63.        }
64.    }
65. }
```

使用 IK Analyzer 内部的智能分词方法对文本文件进行分词处理，将分词后的结果输出到文本文件中，每个词汇以\r\n 换行符进行分割。在本次实验的分词过程中，使用到了拓展词库和停用词库。具体配置方法说明如下：

步骤01 查看配置文件的目录结构，如图 13-22 所示。

图 13-22

步骤02 首先在 src/main/resources 目录下新建文件 XML，选中 src/main/resources 目录右击 File→New→Other，如图 13-23 所示，选择 XML File，单击 Next 按钮。

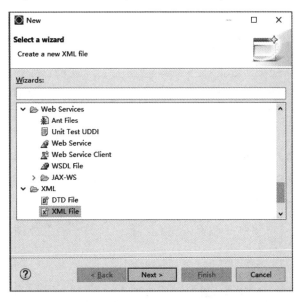

图 13-23

步骤03 将 XML 文件命名为 IKAnalyzer.cfg.xml（在 File name 后的输入框内输入），然后单击 Finish

按钮完成创建，创建完成后打开该 XML 文件，选中 IKAnalyzer.cfg.xml 文件右击 Open With→Text Editor，向 XML 文件中写入如下内容：

```
1.  <?xml version="1.0" encoding="UTF-8"?>
2.  <!DOCTYPE properties SYSTEM "http://java.sun.com/dtd/properties.dtd">
3.  <properties>
4.      <comment>IK Analyzer 扩展配置</comment>
5.      <!--用户可以在这里配置自己的扩展字典 -->
6.      <entry key="ext_dict">ext.dic;</entry>
7.      <!--用户可以在这里配置自己的扩展停止词字典-->
8.      <entry key="ext_stopwords">stopword.dic;</entry>
9.  </properties>
```

XML 文件中指定扩展词和停用词文本的名称。

步骤 04 在 src/main/resources 目录下新建拓展词 File 文件，选中 src/main/resources 目录右击后选择 File→New→Other，如图 13-24 所示，选择 File，单击 Next 按钮。

图 13-24

步骤 05 将 File 文件命名为 ext.dic（在 File name 后的输入框内输入），然后单击 Finish 按钮完成创建，创建完成后打开该 File 文件，选中 ext.dic 文件右击后选择 Open With→Text Editor，向 File 文件中写入想要添加的拓展词汇。如图 13-25 所示。

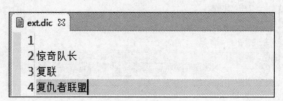

图 13-25

通常情况下分词器默认会把"复"和"联"分开处理，如果我们想让这两个字组合显示则需要向分词器提供这两个字组合的词汇。

步骤06 在 src/main/resources 目录下新建停用词 File 文件，命名为 stopword.dic，向文件中写入想要添加的停用词汇。如图 13-26 所示。

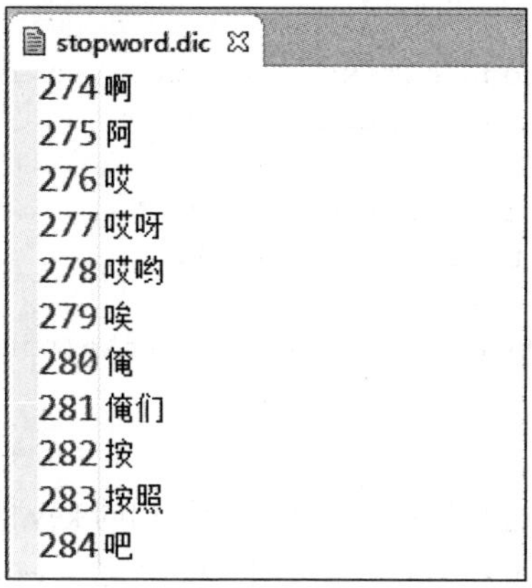

```
stopword.dic
274 啊
275 阿
276 哎
277 哎呀
278 哎哟
279 唉
280 俺
281 俺们
282 按
283 按照
284 吧
```

图 13-26

分词器在分词的过程中会自动去除停用词表中提到的词汇。本项目用到的是哈工大的停用词表，读者也可以根据实际情况自行设计。

13.3.4 第一个 job 的 Map 阶段实现

在 com.mrchi.move 包下新建 Mapper 类 moveMapper，如【代码 13-4】所示。

【代码 13-4】moveMapper.java

```
1.  public class moveMapper extends Mapper<LongWritable, Text, Text, IntWritable> {
2.      private final  IntWritable count= new IntWritable(1);
3.      private Text inkey = new Text();
4.      @Override
5.      protected void map(LongWritable key, Text value, Context context)
6.           throws IOException, InterruptedException {
7.          String word = value.toString();
8.          inkey.set(word);
9.          context.write(inkey,count);
10.     }
11. }
```

Map 阶段获取分词结果文件中的每一个词汇，将这些词汇以 key，value 的形式进行输出，key 为每个词汇，value 为 1。

13.3.5 一个 job 的 Reduce 阶段实现

在 com.mrchi.move 包下新建 Reducer 类 moveReducer，如【代码 13-5】所示。

【代码 13-5】moveReducer.java

```java
1.  public class moveReducer extends Reducer<Text, IntWritable,Text,IntWritable> {
2.
3.      @Override
4.      protected void reduce(Text key, Iterable<IntWritable> values,Context context) throws IOException, InterruptedException {
5.          int total = 0;
6.          for (IntWritable val : values) {
7.              total = total + val.get();
8.          }
9.          context.write( key,new IntWritable(total));
10.     }
11. }
```

将 shuffle 阶段传过来的数据进行处理，shuffle 会将相同 key 的 value 放在一起，reduce 将 value 遍历后聚合，从而得出词汇（key）在所有评论中出现的次数（value）。

13.3.6 第二个 job 的 Map 阶段实现

在 com.mrchi.move 包下新建 Mapper 类 moveSortMapper，如【代码 13-6】所示。

【代码 13-6】moveSortMapper.java

```java
1.  public class moveSortMapper extends Mapper<LongWritable, Text, IntWritable, Text> {
2.      private final static IntWritable wordCount = new IntWritable();
3.      private Text word = new Text();
4.      @Override
5.      protected void map(LongWritable key, Text value, Context context)
6.              throws IOException, InterruptedException {
7.          //以 ":" 作为分隔符处理每一行数据
8.          StringTokenizer tokenizer =
9.                  new StringTokenizer(value.toString(),":");
10.         while (tokenizer.hasMoreTokens()) {
11.             //将分隔后的第一个值赋值给变量 a
12.             String a = tokenizer.nextToken().trim();
13.             word.set(a);
14.             //将分隔后的第二个值赋值给变量 b
15.             String b = tokenizer.nextToken().trim();
16.             //将 b 转成 Integer 类型
17.             wordCount.set(Integer.valueOf(b));
18.             context.write(wordCount, word);
19.         }
20.     }
21. }
```

读取第一个 job 的输出，处理每一行数据使 map 输出为 key、value 形式，key（词汇出现的次数）为 IntWritable 类型，value（词汇）为 Text 类型。

13.3.7　第二个 job 的自定义排序类阶段的实现

在 com.mrchi.move 包下编写自定义排序类 moveSortComparator 继承 WritableComparator，如【代码 13-7】所示。

【代码 13-7】moveSortComparator.java

```
1.  public class moveSortComparator extends WritableComparator{
2.      protected moveSortComparator() {
3.          super(IntWritable.class, true);
4.      }
5.      @Override
6.      public int compare(WritableComparable a, WritableComparable b) {
7.          return -super.compare(a, b);
8.      }
9.  }
```

Reducer 默认排序是从小到大（数字），而我们期望出现次数多的词语排在前面，所以需要重写排序类 WritableComparator。

13.3.8　第二个 job 的自定义分区阶段实现

在 com.mrchi.move 包下编写自定义分区类 moveSortPartition 继承 Partitioner，如【代码 13-8】所示。

【代码 13-8】moveSortPartition.java

```
1.  import org.apache.hadoop.mapreduce.Partitioner;
2.
3.  public class moveSortPartition <K, V> extends Partitioner<K, V> {
4.      @Override
5.      public int getPartition(K key, V value, int numReduceTasks) {
6.          int maxValue = 50;
7.          int keySection = 0;
8.          //只有传过来的 key 值大于 maxValue，并且 numReduceTasks 比如大于 1 个才需要分区，否则直接返回 0
9.          if (numReduceTasks > 1 && key.hashCode() < maxValue) {
10.             int sectionValue = maxValue / (numReduceTasks - 1);
11.             int count = 0;
12.             while ((key.hashCode() - sectionValue * count) > sectionValue) {
13.                 count++;
14.             }
15.             keySection = numReduceTasks - 1 - count;
16.         }
17.         return keySection;
18.     }
19. }
```

如果有多个 Reducer 任务，Reducer 的默认排序只是对发送到该 Reducer 下的数据局部排序。如果想实现全局排序，需要我们手动去写 partitioner。Partitioner 的作用是根据不同的 key，制定相应的规则以分发到不同的 Reducer 中。

13.3.9　第二个 job 的 Reduce 阶段实现

在 com.mrchi.move 包下编写 Reducer 类 moveSortReducer，如【代码 13-9】所示。

【代码 13-9】moveSortReducer.java

```
1.  public class moveSortReducer extends Reducer<IntWritable, Text, IntWritable, Text> {
2.      private Text result = new Text();
3.
4.      @Override
5.      protected void reduce(IntWritable key, Iterable<Text> values,Context context) throws IOException, InterruptedException {
6.          for (Text val : values) {
7.              if(val.toString().length() >1){
8.                  result.set(val.toString());
9.                  context.write(key, result);
10.             }
11.         }
12.     }
13. }
```

将 shuffle 阶段传过来的数据进行处理，遍历相同 key 的 value。

13.3.10　Run 程序主类实现

在 com.mrchi.move 包下编写 Main 类 moveRun，如【代码 13-10】所示。

【代码 13-10】moveRun.java

```
1.  public class moveRun {
2.      public static void main(String[] args) throws IllegalArgumentException, IOException, ClassNotFoundException, InterruptedException {
3.          BasicConfigurator.configure();
4.          Configuration conf = new Configuration();
5.          //定义mapreduce输出key，value的分隔符
6.          conf.set("mapred.textoutputformat.separator", ":");
7.          Job job1 = new Job(conf, "count");
8.          job1.setJarByClass(moveRun.class);
9.          //第一次job的Mapper
10.         job1.setMapperClass(moveMapper.class);
11.         //第一次job的reduce
12.         job1.setReducerClass(moveReducer.class);
13.         //第一次job输出key的文件类型
14.         job1.setOutputKeyClass(Text.class);
15.         //第一次job输出value的文件类型
16.         job1.setOutputValueClass(IntWritable.class);
```

```
17.        //第一次job输入文件的地址
18.        FileInputFormat.addInputPath(job1, new Path("D:\\MoveData\\fencid
ata.txt"));
19.        //第一次job输出结果的地址
20.        FileOutputFormat.setOutputPath(job1, new Path("D:\\MoveData\\resu
lt1"));
21.        job1.waitForCompletion(true);
22.
23.        Job job2 = new Job(conf, "sort");
24.        job2.setJarByClass(moveRun.class);
25.        //map 输出 key 的类型
26.        job2.setMapOutputKeyClass(IntWritable.class);
27.        //map 输出 value 的类型
28.        job2.setMapOutputValueClass(Text.class);
29.        //输出 key 的类型
30.        job2.setOutputKeyClass(IntWritable.class);
31.        //输出 value 的类型
32.        job2.setOutputValueClass(Text.class);
33.        //定义 Mapper 类
34.        job2.setMapperClass(moveSortMapper.class);
35.        //定义 Reducer 类
36.        job2.setReducerClass(moveSortReducer.class);
37.        //自定义排序类
38.        job2.setSortComparatorClass(moveSortComparator.class);
39.        //自定义分区类
40.        job2.setPartitionerClass(moveSortPartition.class);
41.        //输入文件的地址
42.        FileInputFormat.addInputPath(job2,
43. new Path("D:\\MoveData\\result1"));
44.        //输出文件的地址
45.        FileOutputFormat.setOutputPath(job2,
46. new Path("D:\\MoveData\\result2"));
47.        System.exit(job2.waitForCompletion(true) ? 0 : 1);
48.    }
49. }
```

设置 MapReduce 工作任务的相关参数，本实验采用本地运行模式，对指定的本地 D:\\MoveData\\fencidata.txt 目录下的分词结果和第一次 job 的输出 D:\\MoveData\\result1 实现数据清洗，并将结果最终输入到本地 D:\\MoveData\\result2 目录下，设置完毕后，运行程序即可。

13.3.11 编写词云类

在 com.mrchi.move 包下编写词云类 moveShow，如【代码 13-11】所示。

【代码 13-11】moveShow.java

```
1. public abstract class moveShow {
2.     public static void main(String[] args) throws IOException {
3.         BasicConfigurator.configure();
4.         //建立词频分析器，设置词频，以及词语最短长度，此处的参数配置视情况而定即可
```

```
5.         FrequencyAnalyzer frequencyAnalyzer = new FrequencyAnalyzer();
6.         //选取 600 个词
7.         frequencyAnalyzer.setWordFrequenciesToReturn(600);
8.         //引入中文解析器
9.         frequencyAnalyzer.setWordTokenizer(new ChineseWordTokenizer());
10.        //指定文本文件路径,生成词频集合
11.        FrequencyFileLoader ffl = new FrequencyFileLoader();
12.        List<WordFrequency> wordFrequencyList=frequencyAnalyzer
13.            .loadWordFrequencies(ffl.load(new File("D:\\MoveData\\result2\\part-r-00000")));
14.
15.        //设置图片分辨率
16.        Dimension dimension = new Dimension(1920,1080);
17.        //此处的设置采用内置常量即可,生成词云对象
18.        WordCloud wordCloud =
19.            new WordCloud(dimension,CollisionMode.PIXEL_PERFECT);
20.        //设置边界及字体
21.        wordCloud.setPadding(2);
22.        java.awt.Font font = new java.awt.Font("STSong-Light", 2, 20);
23.        //设置词云显示的三种颜色,越靠前设置表示词频越高的词语的颜色
24.        wordCloud.setColorPalette(new LinearGradientColorPalette(Color.RED, Color.BLUE, Color.GREEN, 30, 30));
25.        wordCloud.setKumoFont(new KumoFont(font));
26.        //设置背景色
27.        wordCloud.setBackgroundColor(new Color(255,255,255));
28.        //设置背景图层为圆形
29.        wordCloud.setBackground(new CircleBackground(255));
30.        wordCloud.setFontScalar(new SqrtFontScalar(12, 45));
31.        //生成词云
32.        wordCloud.build(wordFrequencyList);
33.        wordCloud.writeToFile("D:\\MoveData\\move.png");
34.    }
35. }
```

读取 MapReduce 最终输出的结果,选取结果文件中前 600 个词汇组成词云,因为 MapReduce 的第二次 job 做过从大到小的排序,所以选取的会是出现次数较多的前 600 个,根据 key(出现次数)大小在词云中通过词汇字体的大小来体现。

13.3.12 效果测试

为了保证程序正常执行,需要在本地创建 D:\\MoveData\\ 目录,首先执行爬虫类在 D:\\MoveData\\ 目录下会生成 data 的文件。

接着执行分词类在 D:\\MoveData\\ 目录下会生成 fencidata 的文件。

然后执行 moveRun 类在 D:\\MoveData\\ 目录下会生成 result1 和 result2 的两个文件夹。

最后执行 moveShow 类在 D:\\MoveData\\ 目录下会生成 move.png 图片。

最终 D:\\MoveData\\ 目录下的内容如图 13-27 所示。

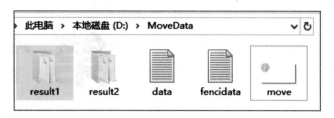

图 13-27

至此，影评分析项目顺利完成，我们可以通过 move 文件查看惊奇队长这部电影的内容。

第 14 章

旅游酒店评价分析项目实战

本章主要内容：

- Hadoop HDFS 数据存储。
- 构建 Hive 数据仓库表。
- 基于 Spark 的数据清洗。
- Hive 数据导出到 MySQL。
- 基于 ECharts 可视化展示。

本章通过一个城市旅游酒店的基本数据及用户评论数据集合，综合运用 Spark 和 Hadoop 进行大数据分析项目实战。项目基于山东省青岛市酒店基本数据和用户评论数据，经过 HDFS 应用程序进行上传和存储，然后基于 HDFS 运用 Spark 进行数据预处理，处理之后的数据构建 Hive 数据仓库，并结合用户需求生成数据仓库分析结果表，将表导出到 MySQL，最后构建基于 Spring Boot 框架的 Web 项目，并结合 ECharts 技术实现数据可视化呈现，为企业和用户提供决策支持。另外，本章涉及 Hadoop 和 Spark 两大框架，均采用伪分布式集群（实际上就是单个虚拟机），服务器节点的 IP 为 192.168.1.110，服务器名称为 oracle，具体环境搭建步骤参见 Hadoop 和 Spark 伪分布式安装对应的章节。

14.1 项目介绍

随着计算机网络发展，各大型网站及平台实时更新，产生了大量的数据。在当今大数据背景下，各行各业积累了海量数据，这些数据具有数据容量大、类型多、增长速度快、价值密度高的特点。许多学者也展开了关于大数据分析算法、分析模式及分析软件工具方面的研究。其中，在大数据结构模型和数据科学理论体系、大数据分析和挖掘基础理论研究方面有很大进步，大数据的应用领域也从科学、工程、电信等领域扩展到各行各业。

在我国，许多规模较大的酒店都有自己的管理系统，提供了完善的酒店管理和酒店预订、评价等服务。部分中小型酒店，由于缺乏资金，依托第三方平台提供在线服务，客户进行操作后，第三方平台会生成记录保存下来。酒店长期积累了大量的在线基本数据和用户评论数据。针对酒店行业，如何利用大数据技术来对现有的数据进行处理和分析，帮助酒店从业者和出行用户提供直观的参考决策，是我们急需解决的问题。一方面，根据用户在线评论数据，帮助酒店从业者提供直观的决策支持，改善酒店管理，以获取最大利润；另一方面，提供某地区的酒店基本满意度情况、酒店分布情况、热门酒店等可视化图表，为用户出行提供可靠参考。

为了使用户对旅游目标城市的酒店住宿和用户满意度、城市各地区酒店分布、用户出游目的等情况有更加直观、明确的了解，并为用户提前规划好住宿和旅游景点的选择提供决策支持，本项目基于山东省青岛市酒店基本数据和用户评论数据构建大数据平台数据仓库，并进行统计分析，最后以 Web 网页的形式将分析结果和决策以可视化图表方式进行展现，也为酒店从业者提供一定的决策支持，方便其在前期市场调查过程中提前了解各区酒店分布、满意度、用户出游目的等。比如，可以根据用户出游类型占比等信息来为酒店从业者规划酒店类型及相关配套。

本项目具体过程为：对青岛市的酒店评论和酒店基本数据进行大数据分析和处理，数据存储到 Hadoop 集群，经过数据清洗后构建 Hive 数据仓库，并基于 Hive 仓库进行数据分析，将分析结果最终导入到 MySQL 中，最后构建基于 JavaEE 的 Web 项目进行酒店数据可视化展示。

本章项目运行前需要搭建 Hadoop 和 Spark 基础环境，这里为了项目方便展开，采用了伪分布式的 Hadoop 集群，服务器的 IP 地址为 192.168.1.110，安装 Hadoop、Hive、Spark 等相关框架或组件。采用 CentOS7 操作系统，MySQL 相关软件安装在该服务器上。开发环境在 Windows 本地主机，使用 IDEA 作为开发工具。

14.2 项目需求及分析

14.2.1 数据集需求

为了给游客提供城市酒店满意度、酒店分布等出行需求，项目对数据集有一定要求。本项目已提供山东省青岛市的酒店基本数据和酒店评论数据两个数据集文件。下面对两个文件的属性做一下说明。

首先是用户评论数据。关于酒店用户评论数据采集的数据格式，描述为：酒店评论的主键（order_id），以便于验证信息的唯一性。酒店网页地址（url），便于查看数据是否采集正确。酒店名字（hotel_name），便于统计酒店数量。酒店评论的发布日期（post_time），便于了解是否是最近发布的。酒店评论的用户名（user_name），便于对评论进行用户画像。本条酒店评论的评论内容（content），这是后面进行酒店顾客意见挖掘的文本数据。酒店评论的用户打分（user_score），这个数据属性能和用户评论进行映射，便于后面的情感分析训练集的制作。

其次是酒店基本数据。关于酒店基本数据采集的数据格式，描述为：酒店 id、酒店名称、酒店评分、评论人数、用户推荐指数（%）、酒店地址、酒店星级、星级详情。

14.2.2 功能需求

本项目对山东省青岛市的酒店数据进行大数据分析和处理。首先需要根据青岛市的酒店基本信息数据和用户评论数据进行清洗，之后上传到 Hadoop 平台的 HDFS 来存储，然后基于 Spark 进行数据清洗，然后基于 HDFS 中的两个数据集构建 Hive 数据仓库；然后基于 Hive 数据仓库，根据用户关心的酒店及评论信息的维度，进行数据分析处理。根据系统功能要求分为 5 个关注的角度，并为每一个关注角度创建 Hive 内部表，具体如下：

（1）用户印象统计，也是用户对该地区总体满意度情况。用户对住过的酒店发表评论同时也可以打分，以下是根据用户对该地区或城市的酒店总体评分情况来统计用户总体印象。评分 4.5~5 分为优良，3.5~4.5 为良好，3.5 以下为差，统计酒店用户评分等级比例。

（2）统计在线评论数最多的十大酒店、十大网络人气酒店。一般情况下，一家酒店的评论数量能代表这家酒店的人气，这里统计的是酒店名称和评论数目。

（3）不同旅游类型占比统计。根据用户评论 Hive 外部表 hotel_data，进行不同旅游类型的统计，根据旅游类型结合用户满意度情况，为用户出行旅游提供参考，了解该地区更适合哪种类型的旅游。

（4）酒店星级分布情况统计。设计酒店星级和数量两个属性，显示不同星级的酒店数量占比，为不同层次的用户提供星级酒店的数量分布。

（5）城市不同地区的酒店数量分布情况。以热力图方式呈现，同时需要显示每个地区酒店数量和平均评论得分情况。

最后，将产生的 Hive 内部表数据，利用 Sqoop 导出到 MySQL 数据库。

数据可视化部分，根据用户 5 个关注的角度的分析结果，构建数据展现的 Web 项目，采用的技术是：SpringBoot+MyBatis+MySQL，开发工具使用 IDEA，图表采用 ECharts 来进行页面图表渲染支持。为了提高页面加载速度和用户体验，采用 Ajax 异步加载的方式来进行图表呈现。

14.3　详细实现

本项目系统的主要数据分析流程如图 14-1 所示。

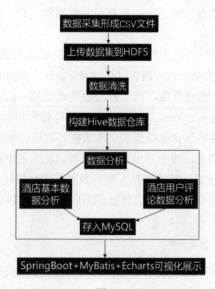

图 14-1

14.3.1 数据集上传到 HDFS

将经过数据预处理后的酒店数据集以及用户评论数据集对应的 CSV 文件上传到 HDFS 存储。这里为了调试代码方便,将程序部分放到 Windows 系统,将 Hadoop 和 Spark 安装到 CentOS7 系统,具体框架安装部分可以参照之前的章节。

步骤01 数据上传程序在 Windows 系统下,打开 IDEA 新建一个 Maven 项目,用于数据上传和数据清洗。具体项目 pom.xml 如【代码 14-1】所示。

【代码 14-1】 pom.xml

```
1.  <!--设置依赖版本号-->
2.  <properties>
3.      <scala.version>2.11.8</scala.version>
4.      <hadoop.version>3.2.2</hadoop.version>
5.      <spark.version>3.1.1</spark.version>
6.  </properties>
7.  <dependencies>
8.      <!-- Scala-->
9.      <dependency>
10.         <groupId>org.scala-lang</groupId>
11.         <artifactId>scala-library</artifactId>
12.         <version>${scala.version}</version>
13.     </dependency>
14.     <!--Spark-->
15.     <dependency>
16.         <groupId>org.apache.spark</groupId>
17.         <artifactId>spark-core_2.11</artifactId>
18.         <version>${spark.version}</version>
19.     </dependency>
20.     <!--Hadoop-->
21.     <dependency>
22.         <groupId>org.apache.hadoop</groupId>
23.         <artifactId>hadoop-client</artifactId>
24.         <version>${hadoop.version}</version>
25.     </dependency>
26.     <dependency>
27.         <groupId>org.apache.hadoop</groupId>
28.         <artifactId>hadoop-common</artifactId>
29.         <version>${hadoop.version}</version>
30.     </dependency>
31.     <dependency>
32.         <groupId>org.apache.hadoop</groupId>
33.         <artifactId>hadoop-hdfs</artifactId>
34.         <version>${hadoop.version}</version>
35.     </dependency>
36.     <dependency>
37.         <groupId>org.apache.hadoop</groupId>
38.         <artifactId>hadoop-mapreduce-client-core</artifactId>
39.         <version>${hadoop.version}</version>
```

```
40.        </dependency>
41.        <dependency>
42.            <groupId>junit</groupId>
43.            <artifactId>junit</artifactId>
44.            <version>4.12</version>
45.        </dependency>
46.        <dependency>
47.            <groupId>org.apache.zookeeper</groupId>
48.            <artifactId>zookeeper</artifactId>
49.            <version>3.4.10</version>
50.        </dependency>
51.        <dependency>
52.                <groupId>com.databricks</groupId>
53.             <artifactId>spark-csv_2.10</artifactId>
54.            <version>1.0.3</version>
55.          </dependency>
56.        <!-- https://mvnrepository.com/artifact/com.google.code.gson/gson
57.        <dependency>
58.            <groupId>com.google.code.gson</groupId>
59.            <artifactId>gson</artifactId>
60.            <version>2.8.0</version>
61.        </dependency>
62.        &lt;!&ndash; https://mvnrepository.com/artifact/org.apache.kafka/kafka &ndash;&gt;
63.        <dependency>
64.            <groupId>org.apache.kafka</groupId>
65.            <artifactId>kafka_2.11</artifactId>
66.            <version>1.0.0</version>
67.        </dependency>-->
68. </dependencies>
```

步骤02 编写具体上传文件的 Java 程序如【代码 14-2】所示。

【代码 14-2】HDFS_CRUD.java

```
1.  import java.io.FileNotFoundException;
2.  import java.io.IOException;
3.  import org.apache.hadoop.conf.Configuration;
4.  import org.apache.hadoop.fs.BlockLocation;
5.  import org.apache.hadoop.fs.FileStatus;
6.  import org.apache.hadoop.fs.FileSystem;
7.  import org.apache.hadoop.fs.LocatedFileStatus;
8.  import org.apache.hadoop.fs.Path;
9.  import org.apache.hadoop.fs.RemoteIterator;
10. import org.junit.Before;
11. import org.junit.Test;
12. public class HDFS_CRUD {
13. FileSystem fs = null;
14. @Before
15. public void init() throws Exception {
16.     //构造一个配置参数对象,设置一个参数:我们要访问的 HDFS 的 URI
```

```
17.         Configuration conf = new Configuration();
18.         //这里指定使用的是 HDFS 文件系统
19.         conf.set("fs.defaultFS", "hdfs://hadoop:9000");
20.         //通过如下的方式进行客户端身份的设置
21.         System.setProperty("HADOOP_USER_NAME", "root");
22.         //通过 FileSystem 的静态方法获取文件系统客户端对象
23.         fs = FileSystem.get(conf);
24.     }
25.     /**
26.      * 将本地的爬取的酒店数据和评论原始数据上传到 HDFS
27.      * @throws IOException
28.      */
29.     @Test
30.     public void testAddFileToHdfs() throws IOException {
31.         //要上传的文件所在本地路径
32.         Path src = new Path("D:/data/hoteldata.csv");
33.         //要上传到 HDFS 的目标路径 文件名
34.         Path dst = new Path("/hdfsdata");
35.         //上传文件方法
36.         fs.copyFromLocalFile(src, dst);
37.         src = new Path("D:/data/hotelbasic.csv");
38.         //要上传到 HDFS 的目标路径 文件名
39.         dst = new Path("/hdfsdata");
40.         //上传文件方法
41.         fs.copyFromLocalFile(src, dst);
42.         //关闭资源
43.         fs.close();
44.     }
45. }
```

项目中已经包含了数据集文件夹，里面有两个文件，hotelbasic.csv 代表酒店基本数据集，hoteldata.csv 代表酒店评论数据集。通过以上程序，把这个两个文件上传到 HDFS 中。

上传后，在 Hadoop 平台使用 fs -ls 命令查看，可以看到多出两个路径，对应评论数据和基本数据，如图 14-2 所示。

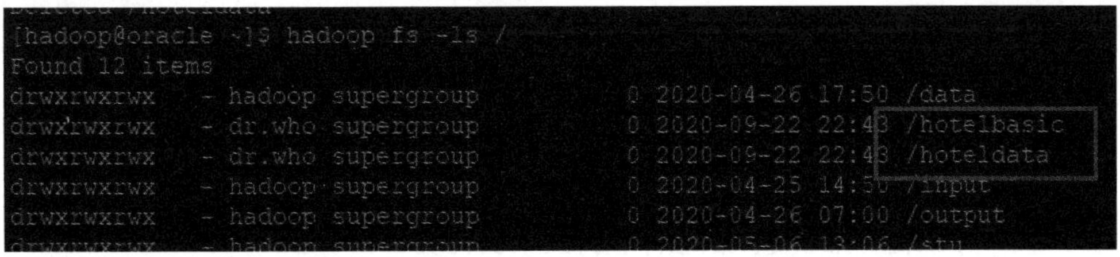

图 14-2

14.3.2 Spark 数据清洗

将上传到 Hadoop 平台的酒店数据进行初步的数据清洗，使其符合大数据分析平台对数据的基本要求。以下是两个数据文件中部分内容：

(1) 部分酒店基本数据属性及原始数据如下：

60169364,枫叶酒店式公寓(青岛金沙滩传媒广场店),4.7,黄岛区珠江路588号传媒广场天相公寓4号楼北侧1号101室。(黄岛金沙滩度假区),hotel_diamond02,经济型
6841087,青岛新天桥快捷宾馆,4.4,市南区肥城路51-1号。(青岛火车站/栈桥/中山路劈柴院),hotel_diamond02,经济型
4640468,欧圣兰廷公寓(青岛万达东方影都店),4.5,黄岛区滨海大道万达公馆A1区2号楼办理入住。(西海岸度假区),hotel_diamond02,经济型
4539535,世纪双帆海景度假酒店(青岛城市阳台店),4.4,黄岛区滨海大道1288号那鲁湾1号楼。(西海岸度假区),hotel_diamond02,经济型
60664769,欧圣兰廷度假公寓(青岛城市阳台店),4.7,黄岛区滨海大道4098号城市阳台风景区世茂悦海13号楼一楼。(西海岸度假区),hotel_diamond02,经济型
823437,青岛花园大酒店—贵宾楼,4.8,市南区彰化路6号贵宾楼。(五四广场/万象城/奥帆中心/市政府青岛大学),hotel_diamond04,高档型
43679094,青岛燕岛之星度假公寓,4.7,市南区燕儿岛路15号1楼大厅。(五四广场/万象城/奥帆中心/市政府),hotel_diamond02,经济型
22416789,容锦酒店(青岛台东步行街店),4.6,市北区延安路129号利群百惠商厦6-7楼。(台东步行街/啤酒街),hotel_diamond02,经济型
17263343,慢居听海酒店(青岛吾悦广场店),4.8,黄岛区滨海大道2888号梦时代广场17号楼1楼大厅101室。(黄岛金沙滩度假区),hotel_diamond02,经济型
427962,安澜宾舍酒店(青岛东海中路海滨店),4.4,市南区东海中路30号银海大世界院内。(五四广场/万象城/奥帆中心/市政府),hotel_diamond02,经济型
21122269,青岛蓝朵海景假日公寓,4.8,崂山区秦岭路19号协信中心3号楼28层。(国际会展中心/石老人海水浴场),经济型

(2) 部分酒店用户评论属性及原始数据如下（评论内容列已去掉）：

Elainemimi,枫叶酒店式公寓(青岛金沙滩传媒广场店),其他,20-Jul,2020/7/6,5
_WeChat268192****,枫叶酒店式公寓(青岛金沙滩传媒广场店),商务出差,20-Jun,2020/6/28,5
_CFT010000002415****,枫叶酒店式公寓(青岛金沙滩传媒广场店),朋友出游,20-Jun,2020/6/26,5
_WeChat320342****,枫叶酒店式公寓(青岛金沙滩传媒广场店),情侣出游,20-Jun,2020/6/20,5
_WeChat320342****,枫叶酒店式公寓(青岛金沙滩传媒广场店),情侣出游,20-Jun,2020/6/21,5
_WeChat320342****,枫叶酒店式公寓(青岛金沙滩传媒广场店),情侣出游,20-Jun,2020/6/21,5
M24589****,枫叶酒店式公寓(青岛金沙滩传媒广场店),独自旅行,20-May,2020/6/4,5
M221656****,青岛新天桥快捷宾馆,家庭亲子,20-Aug,2020/8/27,5
M419190****,青岛新天桥快捷宾馆,家庭亲子,20-Jul,2020/8/10,5
M354977****,青岛新天桥快捷宾馆,独自旅行,20-Jul,2020/7/30,5
M355264****,青岛新天桥快捷宾馆,独自旅行,20-Jul,2020/8/3,5
M311789****,青岛新天桥快捷宾馆,其他,20-Jul,2020/8/3,5
朝生牧者,青岛新天桥快捷宾馆,独自旅行,20-Jul,2020/8/10,5
_WeChat229580****,青岛新天桥快捷宾馆,情侣出游,20-Aug,2020/8/26,4.5
M415985****,青岛新天桥快捷宾馆,独自旅行,20-Jul,2020/8/6,5
_WeChat385461****,青岛新天桥快捷宾馆,朋友出游,20-Aug,2020/8/7,3.8
M416989****,青岛新天桥快捷宾馆,商务出差,20-Jul,2020/8/10,5
M272306****,青岛新天桥快捷宾馆,情侣出游,20-Jun,2020/6/22,5
w风清,青岛新天桥快捷宾馆,情侣出游,20-Jun,2020/8/10,5
_WeChat290262****,青岛新天桥快捷宾馆,朋友出游,20-Jun,2020/8/10,5
杰科1105,青岛新天桥快捷宾馆,家庭亲子,20-Jul,2020/7/22,5
M317651****,青岛新天桥快捷宾馆,家庭亲子,20-Aug,2020/8/18,3.8
通帕蓬廖化,青岛新天桥快捷宾馆,独自旅行,20-Aug,2020/8/19,4.8

```
M327775****,青岛新天桥快捷宾馆,朋友出游,20-Aug,2020/8/11,5
_WeChat270171****,青岛新天桥快捷宾馆,家庭亲子,20-Aug,2020/8/6,5
平哥儿走天涯,青岛新天桥快捷宾馆,商务出差,20-Jul,2020/8/7,5
_WeChat266127****,青岛新天桥快捷宾馆,朋友出游,20-Aug,2020/8/12,5
```

主要工作：

（1）酒店基本数据集中的酒店星级类型这一列数据叫法不一致，比如有的酒店叫四星级，有的则叫高档型，这里统一做一下处理，将所有的"国家旅游局评定为四星级"替换为"高档型"，将"国家旅游局评定为三星级"替换为"舒适型"，将"国家旅游局评定为二星级"替换为"经济型"，将"国家旅游局评定为五星级"替换为"豪华型"。

（2）由于大数据服务平台这个子模块并不对用户具体评论内容进行情感分析，情感分析交给情感分析子系统来处理，所以这里将评论内容数据列去掉。

（3）删除所有空行。

（4）从酒店地址中提取区县名称，替换掉地址那一列内容，为区县酒店分布统计提供标准数据。

具体清洗代码用 Scala 程序编写，具体如【代码 14-3】所示。

【代码 14-3】SparkProcessSubmit.scala

```scala
1.  import org.apache.spark.sql.functions.udf
2.  import org.apache.spark.sql.{DataFrame, SQLContext, SaveMode}
3.  import org.apache.spark.{SparkConf, SparkContext}
4.
5.  //编写单词技术
6.  object SparkProcessSubmit {
7.    def fc(a: String):String = {
8.      if(a.contains("平度")) return "平度市";
9.      else if(a.contains("市南")) return "市南区"
10.     else if(a.contains("市北")) return "市北区"
11.     else if(a.contains("李沧")) return "李沧区"
12.     else if(a.contains("城阳")) return "城阳区"
13.     else if(a.contains("崂山")) return "崂山区"
14.     else if(a.contains("黄岛")) return "黄岛区"
15.     else if(a.contains("即墨")) return "即墨区"
16.     else if(a.contains("胶州")) return "胶州市"
17.     else if(a.contains("莱西")) return "莱西市"
18.     else return "其他";
19.   }
20.   def main(args: Array[String]): Unit = {
21.     val input = "hdfs://192.168.1.110:9000/"
22.     val output =""
23.     val appName = "etl"
24.     val master = "spark://192.168.1.110:7077"
25.     val conf = new SparkConf().setAppName(appName).setMaster(master).set("spark.executor.memory","0.5g")//每个 executor 的内存
26.     val sc = new SparkContext(conf)
27.     val sqlContext = new SQLContext(sc)
```

```
28.         val df: DataFrame = sqlContext.read.format("com.databricks.spark.c
sv").option("header", "true") //在 csv 第一行有属性"true"，没有就是"false"
29.             .option("inferSchema", true.toString)//这是自动推断属性列的数据类型
30.             .load(input+"hdfsdata/hotelbasic.csv")
31.         val deffun = udf(fc(_:String))
32.         import org.apache.spark.sql.functions//如何在 DataFrame 上使用 Spark 中
DataFrameNaFunctions 类提供的函数，须调用 na
33.         df.dropDuplicates().na.drop(how="all").na.replace("星级详情",Map("
国家旅游局评定为四星级"->"高档型","国家旅游局评定为五星级"->"豪华型","国家旅游局评定为三星级
"->"舒适型","国家旅游局评定为二星级"->"经济型"))
34.             .withColumn("酒店地址", deffun(functions.col("酒店地址")))
35.             .coalesce(1).write.mode(SaveMode.Overwrite).option("mapreduce.
fileoutputcommitter.marksuccessfuljobs","false")
36.             .option("header","false").option("sep",",").csv(input+"hotelba
sic")
37.         //clean hoteldata//////////////////
38.         val df1= sqlContext.read.format("com.databricks.spark.csv").option("
header", "true") //在 csv 第一行有属性"true"，没有就是"false"
39.             .option("inferSchema", true.toString)//这是自动推断属性列的数据类型
40.             .load(input+"hdfsdata/hoteldata.csv")
41.         df1.drop("content").select("user_name","hotel_name","type","craw_ti
me","post_time","user_score")
42.             .dropDuplicates()//删除重复行
43.             .na.drop("all")
44.             .coalesce(1).write.mode(SaveMode.Overwrite).option("mapreduce.
fileoutputcommitter.marksuccessfuljobs","false")
45.             .option("header","false").option("sep",",").csv(input+"hotelda
ta")
46.         //8.关闭 sparkContext 对象
47.         sc.stop()
48.     }
49. }
```

清洗后的数据分别存放到 HDFS 的两个不同目录，hotelbasic 存放基本数据，hoteldata 存放评论数据，如图 14-3 所示。

图 14-3

14.3.3 构建 Hive 数据仓库表

酒店大数据分析服务平台，创建了两个外部表，分别对应酒店基本信息表和酒店用户评论表。

Hive 内部表是分别对 5 个关于酒店数据的方面进行分析得到的结果，最终内部表数据会导入到 MySQL 中。

各表结构设计如表 14-1~表 14-5 所示。

表14-1 出游类型统计表设计

字段编号	字段名称	数据类型	约束	字段描述
1	Triptype	varchar	PK	出游类型
2	num	int		数量

表14-2 用户满意度分布统计表设计

字段编号	字段名称	数据类型	约束	不是 Nulls	字段描述
1	bad	int		☑	不满意
2	good	int		☑	满意
3	excellent	int		☑	非常满意

表14-3 酒店星级分布统计表

字段编号	字段名称	数据类型	约束	不是 Nulls	字段描述
1	Stardetail	varchar	PK	☑	酒店星级类型
2	nums	Int		☑	数量

表14-4 酒店评论数据数量统计表

字段编号	字段名称	数据类型	约束	字段描述
1	hotel_name	varchar	PK	酒店星期
2	num	int		数量

表14-5 各区县酒店数量和用户推荐评分设计

字段编号	字段名称	数据类型	约束	字段描述
1	Area_name	varchar	PK	区县名称
2	num	int		数量
3	recommend	Int		推荐平均评分

1. 基于上传的酒店用户评论数据 hoteldata.csv 创建 Hive 外部表

```
create external table hotel_data(user_name string,hotel_name string,trip_typ
e string,time1 string,time2 string ,user_score double)
    ROW FORMAT DELIMITED FIELDS TERMINATED BY ',' LOCATION '/hoteldata';
```

这样数据就会自动加载/hoteldata 下面的数据。

2. 基于上传的酒店基本数据 hotelbasic.csv 创建 Hive 表

```
create external table hotel_basic(id string,name string,score double,comment
num int,recommend int,address string,star  string,stardetail string)
    ROW FORMAT DELIMITED FIELDS TERMINATED BY ',' LOCATION '/hotelbasic';
```

这样数据就会自动加载/hotelbasic 下面的数据。

3. 基于 Hive 外部表统计以下数据形成内部表,并将分析结果导出到 MySQL 数据库

(1) 用户印象统计,用户住过酒店发表评论同时也可以打分,以下是根据用户对该地区或城市的酒店总体评分情况来统计用户总体印象。评分 4.5~5 分为优良,3.5~4.5 为良好,3.5 以下为差,

统计酒店用户评分等级比例。

提前建好 Hive 内部表 score_stat 表结构：

```
create table score_stat(bad int,good int,excellent int)
```

覆盖式插入数据：

```
insert overwrite table score_stat select  count(case when user_score<3.5 then 1 else null end) as 'bad',count(case when user_score>=3.5 and user_score<4.5 then 1 else null end) as 'good',
    count(case when user_score>=4.5 then 1 else null end) as 'excellent'  from hotel_data
```

将数据导出到 MySQL 中：

```
sqoop export --connect jdbc:mysql://192.168.1.110:3306/test --username root --password root --table score_stat --fields-terminated-by '\001' --export-dir '/user/hive/warehouse/score_stat '
```

（2）统计在线评论数最多的十大酒店，十大网络人气酒店。

一般情况下，一家酒店的评论数量能代表这家酒店的人气，这里统计的是酒店名称和评论数目。

排名前十的酒店，设计评论数目表结构：

```
create table comments_stat(hotel_name string,nums int)
insert overwritetable comments_stat select hotel_name,count(1) as nums from hotel_data group by hotel_name
```

将该数据导入到 MySQL，Sqoop 导入命令：

```
sqoop export --connect jdbc:mysql://192.168.1.110:3306/test --username root --password root --table comments_stat --fields-terminated-by '\001' --export-dir '/user/hive/warehouse/ comments_stat '
```

（3）不同旅游类型占比统计。

```
create table triptype_stat(triptype string,nums bigint)
insert overwrite table triptype_stat  select trip_type,count(1) as nums from hotel_data group by trip_type
```

设计 MySQL 表结构：

```
create table triptype_stat(triptype varchar(33),nums int)
```

将该数据导入到 MySQL，Sqoop 导入命令：

```
sqoop export --connect jdbc:mysql://192.168.1.110:3306/test --username root --password root --table triptype_stat --fields-terminated-by '\001' --export-dir '/user/hive/warehouse/triptype_stat '
```

（4）酒店星级分布情况统计，设计酒店星级和数量结构，如下：

```
create table star_stat(stardetail string,nums int)
insert overwrite table star_stat  select stardetail,count(1) from hotel_basic group by stardetail
```

设计 MySQL 表结构:

```
create table star_stat(stardetail varchar(33),nums int)
```

将该数据导入到 MySQL,Sqoop 导入命令:

```
sqoop export --connect jdbc:mysql://192.168.1.110:3306/test --username root --password root --table star_stat --fields-terminated-by '\001' --export-dir '/user/hive/warehouse/star_stat '
```

(5) 城市不同区的酒店数量分布情况,以热力图的方式呈现。

```
create table area_stat(area_name string,nums int,recommend int)
```

插入数据:

```
insert overwrite table area_stat  select address,round(avg(recommend),2)  from hotel_basic group by regexp_extract(address,'(.+?区|市|县)(.*)',1);
```

以上查询语句是使用正则表达式从酒店数据的酒店地址中提取区县的名称。

设计 MySQL 表结构:

```
create table area_stat (area_name varchar(33),nums int,recommend int)
```

将该数据导入到 MySQL,Sqoop 导入命令:

```
sqoop export --connect jdbc:mysql://192.168.1.110:3306/test --username root --password root --table area_stat --fields-terminated-by '\001' --export-dir '/user/hive/warehouse/area_stat '
```

将以上命令形成 shell 脚本,命名为 bigdata.sh,通过运行脚本来构建数据仓库并取得分析结果。运行完整代码如下。

【代码 14-4】bigdata.sh

```
1. #!/bin/sh
2. #创建外部表
3. echo "基于上传的 hoteldata.csv 创建 hive 外部表"
4. hive -e "create external table if not exists  hotel_data(user_name string,hotel_name string,trip_type string,time1 string,time2 string ,user_score double)
5. ROW FORMAT DELIMITED FIELDS TERMINATED BY ',' LOCATION '/hoteldata'"
6. echo "基于上传的 hotelbasic.csv 创建 hive 表"
7. hive -e "create external table if not exists  hotel_basic(id string,name string,score double,commentnum int,recommend int,address string,star string,stardetail string)
8. ROW FORMAT DELIMITED FIELDS TERMINATED BY ',' LOCATION '/hotelbasic'"
9. #创建 hive 内部表
10. echo "======================创建 hive 内部表======================"
11. #1.来该地区用户的旅游类型统计
12. echo "1.用户的旅游类型统计表......................................."
13. hive -e "create table triptype_stat(triptype string,nums bigint)"
14. hive -e "insert overwrite table triptype_stat  select trip_type,count(1) as nums from hotel_data group by trip_type"
15. #将数据导出到 MySQL
16. echo "导入数据到 mysql 开始"
```

```
17.  sqoop export --connect 
jdbc:mysql://192.168.1.110:3306/hotel?useUnicode=true\&characterEncoding=utf-8 
--username root --password 123456 --table triptype_stat --fields-terminated-by 
'\001' --export-dir '/user/hive/warehouse/triptype_stat'
18.  echo "导入数据到 mysql 结束"
19.  hive -e "drop table  triptype_stat;"
20.  echo "导入 mysql 之后删除 hive 内部表 
triptype_stat........................................."
21.  #2.根据用户对该地区或城市的酒店总体评分情况来统计用户总体印象, 用户评分统计
22.  echo "2.用户对该地区或城市的酒店总体评分情况来统计用户总体印象..."
23.  hive -e "create table score_stat(bad int,good int,excellent int)"
24.  hive -e "insert overwrite table score_stat select  count(case when 
user_score<3.5 then 1 else null end)  as bad,count(case when user_score>=3.5 and 
user_score<4.5 then 1 else null end)  as good,count(case when user_score>=4.5 then 
1 else null end)  as excellent  from hotel_data"
25.  #将数据导出到 MySQL
26.  echo "导入数据到 mysql 开始"
27.  sqoop export --connect 
jdbc:mysql://192.168.1.110:3306/hotel?useUnicode=true\&characterEncoding=utf-8 
--username root --password 123456 --table score_stat --fields-terminated-by '\001' 
--export-dir '/user/hive/warehouse/score_stat'
28.  echo "导入数据到 mysql 结束"
29.  hive -e "drop table  score_stat;"
30.  echo "导入 mysql 之后删除 hive 内部表 score_stat........"
31.  #3.十大网络人气酒店, 酒店名称和用户评论数量
32.  echo "3.十大网络人气酒店........................................"
33.  hive -e "create table comments_stat(hotel_name string,nums int)"
34.  hive -e "insert overwrite table comments_stat select hotel_name,count(1) 
as nums from hotel_data group by hotel_name"
35.  #将数据导出到 MySQL
36.  echo "导入数据到 mysql 开始"
37.  sqoop export --connect 
jdbc:mysql://192.168.1.110:3306/hotel?useUnicode=true\&characterEncoding=utf-8 
--username root --password 123456 --table comments_stat --fields-terminated-by 
'\001' --export-dir '/user/hive/warehouse/comments_stat'
38.  echo "导入数据到 mysql 结束"
39.  hive -e "drop table  comments_stat;"
40.  echo "导入 mysql 之后删除 hive 内部表 comments_stat......................"
41.  #4.酒店星级分布情况统计, 设计酒店星级和数量结构
42.  echo "4.酒店星级分布情况统计............................"
43.  hive -e "create table star_stat(stardetail string,nums int)"
44.  hive -e "insert overwrite table star_stat  select stardetail,count(1) from 
hotel_basic group by stardetail"
45.  #将数据导出到 MySQL
46.  echo "导入数据到 mysql 开始"
47.  sqoop export --connect 
jdbc:mysql://192.168.1.110:3306/hotel?useUnicode=true\&characterEncoding=utf-8 
--username root --password 123456 --table star_stat  --fields-terminated-by '\001' 
--export-dir '/user/hive/warehouse/star_stat'
48.  echo "导入数据到 mysql 结束"
49.  hive -e "drop table  star_stat;"
50.  echo "导入 mysql 之后删除 hive 内部表 star_stat................"
51.  #5.各地区酒店数量统计
52.  echo "各地区酒店数量统计......................................"
53.  hive -e "create table area_stat(area_name string,nums int,recommend int)"
```

```
54. hive -e "insert overwrite table area_stat  select
address ,count(1),round(avg(recommend),2)  from hotel_basic group by  address"
55. #将数据导出到 MySQL
56. echo"导入数据到mysql 开始"
57. sqoop export --connect
jdbc:mysql://192.168.1.110:3306/hotel?useUnicode=true\&characterEncoding=utf-8
--username root --password 123456 --table area_stat  --fields-terminated-by '\001'
--export-dir '/user/hive/warehouse/area_stat'
58. echo "导入数据到mysql 结束"
59. hive -e "drop table  area_stat;"
60. echo"导入mysql 之后删除hive 内部表area_stat..............................."
```

运行过程如图 14-4 所示。

图 14-4

Hive 通过以上脚本的运行，根据 Hive 仓库外部表数据，进行城市酒店数据评分、用户印象、各区县酒店数量和评分、网络人气、酒店星级、来此地游客的旅游目的类型统计等数据分析，每一项统计都构建 Hive 内部表。

MySQL 数据库对应的表结构需要提前创建好，以方便 Hive 将内部表数据导出到 MySQL，建表脚本如下：

```
create table triptype_stat(triptype varchar(33),nums int);
create table score_stat(bad int,good int,excellent int);
create table comments_stat(hotel_name varchar(200),nums int);
```

```
create table star_stat(stardetail varchar(200),nums int)
create table area_stat(area_name varchar(200),nums int,recommend int)
```

Hive 表数据导出到 MySQL 需要用到 Sqoop 组件。

14.3.4　Hive 表数据导出到 MySQL

Sqoop 是一个数据迁移工具。Sqoop 非常简单，其整合了 Hive、HBase 和 Oozie，通过 MapReduce 任务来传输数据，从而提供并发特性和容错。Sqoop 由于是将数据导入到 HDFS 中，所以需要依赖于 Hadoop，即前提是 Hadoop 已经安装且正确配置。

Sqoop 主要通过 JDBC 和关系数据库进行交互。理论上支持 JDBC 的 DataBase 都可以使用 Sqoop 和 HDFS 进行数据交互。比如将 DataBase 中的数据导入到 HDFS 或是将 HDFS 中的数据导入到 DataBase 中。

1. import

import 命令，用于将数据库中的数据导入到 HDFS。其中--table 参数用于将一个表中的数据全部的导入到 HDFS 中去。

（1）--table 指定导出的表

```
#!/bin/bash
./sqoop import \
--connect \
jdbc:mysql://192.168.1.110:3306/opt?characterEncoding=UTF-8 \
--username root \
--password 1234 \
--table studs \          #指定表名
-m 2 \                   #指定mapper的个数，不能超过集群节点的数量，默认为4
--split-by "id" \        #只要-m不是1必须指定分组的字段名称
--where "age>100 and sex='1'" \   #指定where条件，可以使用""双引号
--target-dir /out001     #指定导入到HDFS以后目录
```

默认导出到 HDFS 的数据以"，"（逗号）分开如下所示。

```
$ hdfs dfs -cat /out001/*
2a56b3536b544f289ba79b2b5c1196c4,Jerry,89e4a3e..3ee3e03946d85d
cc645dc7811740fc9856b1c7c8e19e89,Alex,c924e3..5a0487e23207986189d
U001,Jack,1234
U002,Mike,1234
```

可以使用--fields-terminated-by 参数，指定分隔符号，如--fields-terminated-by "\t"将分隔符号设置为制表符。

（2）--query 指定查询语句

如果在 import 中已经使用了--query 语句，则--where 和--table 将被忽略。在--query 所指定的语句中，必须将$CONDITIONS 作为条件添加到 where 子句中。如果--query 使用后面使用""""（双引号），则应该使用\$CONDITIONS。注意前面的"\"（斜线）。

```
#!/bin/bash
```

```
sqoop import \
--connect jdbc:mysql://192.168.1.110:3306/qlu?characterEncoding=UTF-8 \
--username root \
--password 1234 \
#注意以下使用的 SQL 语句,如果使用""(双引号)则必须添加\在$CONDITIONS 前面
--query "select name,sex,age,addr from studs where sex='0' and addr like '山
东%' and \$CONDITIONS" \
--split-by "name" \
--fields-terminated-by "\t" \   #使用制表符号进行数据分隔
--target-dir /out002 \
-m 2
```

--query 参数的 SQL 可以写得很复杂,比如下面的示例是一个关联的查询语句:

```
#!/bin/bash
sqoop import \
--connect jdbc:mysql://192.168.1.110:3306/studs?characterEncoding=UTF-8 \
--username root \
--password 1234 \
--query \
"select s.stud_id,s.stud_name as sname,c.course_name as cname \
from studs s inner join sc on s.stud_id=sc.sid \
   inner join courses c on c.course_id=sc.cid where \$CONDITIONS" \
--target-dir /out004 \
--split-by "s.stud_id" \    #根据某个列进行分组
-m 2
```

2. export 导出到关系型数据库中去

使用 sqoop export 命令,可以将 HDFS 数据导出到关系型数据库中去。

```
#!/bin/bash
sqoop export \
--connect \
 jdbc:mysql://192.168.1.110:3306/opt?characterEncoding=UTF-8 \
--username root \
--password 1234 \
--export-dir /out001 \ #指定导出的目录
--table "studs" \ #指定 HDFS 中数据与数据库中表列的对应关系
--columns "stud_id,stud_age,stud_name" \#指定 HDFS 中数据进行分隔
--fields-terminated-by "\t" \
-m 2
```

14.3.5 数据可视化开发

将 14.3.3 节中的分析结果，即 MySQL 中的数据以图表的方式呈现到 Web 网页上。

这里采用 Spring Boot+MyBatis+MySQL 技术构建可视化项目，本节具体代码可以参见本章配套代码文件夹下的 HotelVisualization。

图表展示使用 ECharts，ECharts 是一个纯 JavaScript 图表库，底层依赖于轻量级的 Canvas 类库 ZRender，基于 BSD 开源协议，是一款非常优秀的可视化前端框架。它提供直观、生动、可交互、可高度个性化定制的数据可视化图表。创新的拖曳重计算、数据视图、值域漫游等特性大大增强了用户体验，赋予了用户对数据进行挖掘、整合的能力。支持折线图（区域图）、柱状图（条状图）、散点图（气泡图）、K 线图、饼图（环形图）等。

首先，新建一个 Maven 项目，引入需要使用的框架，如 SpringBoot、MyBatis 等，具体 pom.xml 代码如【代码 14-5】所示。

【代码 14-5】pom.xml

```
1.  <?xml version="1.0" encoding="UTF-8"?>
2.  <project xmlns="http://maven.apache.org/POM/4.0.0" xmlns:xsi="http://www.w3.org/2001/XMLSchema-instance"
3.        xsi:schemaLocation="http://maven.apache.org/POM/4.0.0 http://maven.apache.org/xsd/maven-4.0.0.xsd">
4.      <modelVersion>4.0.0</modelVersion>
5.      <parent>
6.          <groupId>org.springframework.boot</groupId>
7.          <artifactId>spring-boot-starter-parent</artifactId>
8.          <version>2.1.6.RELEASE</version>
9.          <relativePath/> <!-- lookup parent from repository -->
10.     </parent>
11.     <groupId>com.zjp.echartsdemo</groupId>
12.     <artifactId>echartsdemo</artifactId>
13.     <version>0.0.1-SNAPSHOT</version>
14.     <name>echartsdemo</name>
15.     <description>Demo project for Spring Boot</description>
16.     <properties>
17.         <java.version>1.8</java.version>
18.     </properties>
19.     <dependencies>
20.         <dependency>
21.             <groupId>org.springframework.boot</groupId>
22.             <artifactId>spring-boot-starter-web</artifactId>
23.         </dependency>
24.         <dependency>
25.             <groupId>org.springframework.boot</groupId>
26.             <artifactId>spring-boot-starter-test</artifactId>
27.             <scope>test</scope>
28.         </dependency>
29.         <dependency>
30.             <groupId>org.webjars.bower</groupId>
31.             <artifactId>echarts</artifactId>
```

```xml
32.            <version>4.2.1</version>
33.        </dependency>
34.        <dependency>
35.            <groupId>org.webjars</groupId>
36.            <artifactId>jquery</artifactId>
37.            <version>3.4.1</version>
38.        </dependency>
39.        <dependency>
40.            <groupId>org.mybatis.spring.boot</groupId>
41.            <artifactId>mybatis-spring-boot-starter</artifactId>
42.            <version>2.0.1</version>
43.        </dependency>
44.        <dependency>
45.            <groupId>mysql</groupId>
46.            <artifactId>mysql-connector-java</artifactId>
47.            <scope>runtime</scope>
48.        </dependency>
49.        <dependency>
50.            <groupId>com.github.pagehelper</groupId>
51.            <artifactId>pagehelper-spring-boot-starter</artifactId>
52.            <version>1.2.5</version>
53.        </dependency>
54.        <!-- alibaba 的 druid 数据库连接池 -->
55.        <dependency>
56.            <groupId>com.alibaba</groupId>
57.            <artifactId>druid-spring-boot-starter</artifactId>
58.            <version>1.1.9</version>
59.        </dependency>
60.        <!-- 引入 Thymeleaf 依赖 -->
61.        <dependency>
62.            <groupId>org.springframework.boot</groupId>
63.            <artifactId>spring-boot-starter-thymeleaf</artifactId>
64.        </dependency>
65.        <dependency>
66.            <groupId>org.springframework.boot</groupId>
67.            <artifactId>spring-boot-devtools</artifactId>
68.            <optional>true</optional>
68.        </dependency>
70.    </dependencies>
71.    <build>
72.        <plugins>
73.            <plugin>
74.                <groupId>org.springframework.boot</groupId>
75.                <artifactId>spring-boot-maven-plugin</artifactId>
76.            </plugin>
77.        </plugins>
78.    </build>
79. </project>
```

然后，配置 SpringBoot 的配置文件，定义数据库连接池配置以及 Web 服务器端口配置，具体如【代码 14-6】所示。

【代码 14-6】application.properties

```
1.  server.port=8088
2.  #数据库连接池配置
3.  spring.datasource.name=zjptest
4.  spring.datasource.type=com.alibaba.druid.pool.DruidDataSource
5.  spring.datasource.druid.filters=stat
6.  spring.datasource.druid.driver-class-name=com.mysql.jdbc.Driver
7.  spring.datasource.druid.url=jdbc:mysql://192.168.1.110:3306/hotel?useUnicode=true&characterEncoding=UTF-8&allowMultiQueries=true&serverTimezone=UTC
8.  spring.datasource.druid.username=root
9.  spring.datasource.druid.password=123456
10. pring.datasource.druid.initial-size=1
11. spring.datasource.druid.min-idle=1
12. spring.datasource.druid.max-active=20
13. spring.datasource.druid.max-wait=6000
14. spring.datasource.druid.time-between-eviction-runs-millis=60000
15. spring.datasource.druid.min-evictable-idle-time-millis=300000
16. spring.datasource.druid.validation-query=SELECT 'x'
17. spring.datasource.druid.test-while-idle=true
18. spring.datasource.druid.test-on-borrow=false
19. spring.datasource.druid.test-on-return=false
20. spring.datasource.druid.pool-prepared-statements=false
21. spring.datasource.druid.max-pool-prepared-statement-per-connection-size=20
22. #mybatis配置
23. mybatis.mapper-locations=classpath:mapper/*.xml
24. mybatis.type-aliases-package= com.dpzhou.echartsdemo.echartsdemo.entity
25. #thymeleaf配置
26. spring.thymeleaf.cache=false
27. spring.thymeleaf.prefix=classpath:/templates/
28. spring.thymeleaf.suffix=.html
29. spring.thymeleaf.mode=HTML5
30. spring.thymeleaf.encoding=UTF-8
31. spring.thymeleaf.check-template-location=true
```

根据项目需求，需要开发 5 个可视化统计图，因为项目使用了 **MyBatis** 框架，所以可以将 **SQL** 直接配置在 **Mapper** 文件中，详细如【代码 14-7】所示。

【代码 14-7】CommonMapper.xml

```
1.  <?xml version="1.0" encoding="UTF-8"?>
2.  <!DOCTYPE mapper PUBLIC "-//mybatis.org//DTD Mapper 3.0//EN" "http://mybatis.org/dtd/mybatis-3-mapper.dtd">
3.  <mapper namespace="com.zjp.echartsdemo.echartsdemo.dao.CommonMapper">
4.      <select id="selectTripType" parameterType="java.lang.String" resultType="map">
5.          select
6.          triptype,nums
7.          from triptype_stat order by nums desc limit 0,7
```

```
8.      </select>
9.      <select id="selectCommentsStat" parameterType="java.lang.String" resultType="map">
10.         select
11.             hotel_name,nums
12.         from comments_stat order by nums desc limit 0,10
13.     </select>
14.     <select id="selectScoreStat" parameterType="java.lang.String" resultType="map">
15.         select
16.             *
17.         from score_stat
18.     </select>
19.     <select id="selectStarStat" parameterType="java.lang.String" resultType="map">
20.         select
21.             stardetail,nums
22.         from star_stat order by nums desc limit 0,4
23.     </select>
24.     <select id="selectAreaStat" parameterType="java.lang.String" resultType="map">
25.         select * from area_stat where area_name in('市南区','市北区','黄岛区','即墨区','城阳区','崂山区','李沧区') order by nums desc limit 0,10
26.     </select>
27. </mapper>
```

前端异步请求数据采用 Ajax 技术，后台查询数据以 JSON 格式返回，通过 ECharts 在页面加载返回 JSON 数据。

定义 view.html 页面，划分 DIV 分别显示对应的统计图表，页面 JS 脚本采用 Ajax 实现无刷新交互，具体页面如【代码 14-8】所示。

【代码 14-8】view.html

```
1.  <!-- 为 ECharts 准备一个具备大小（宽、高）的 DOM -->
2.  <h1 class="aa">青岛市酒店大数据可视化</h1>
3.  <div id="main" style="width: 500px;height:400px;position:absolute;top:100px"></div><!--用户的旅游类型统计-->
4.  <div id="main2" style="width: 500px;height:400px;position:absolute;top:100px;left:550px"></div><!--地区酒店数量统计-->
5.  <div id="main3" style="width: 500px;height:400px;position:absolute;top:100px;left:1050px"></div><!--用户印象评分等级比例统计-->
6.  <a href="/echarts" >Goto 基于用户评论统计分析可视化</a>
7.  <script type="text/javascript">
8.      //基于准备好的 dom，初始化 ECharts 实例
9.      var myChart = echarts.init(document.getElementById('main'));
10.     //新建 productName 与 nums 数组来接收数据
11.     var triptypes = [];
12.     var nums = [];
13.     var json = {};
14.     var datatemp = [];
```

```javascript
15.    //旅游类型统计
16.    $.ajax({
17.        type:"GET",
18.        url:"/triptypestat",
19.        dataType:"json",
20.        async:false,
21.        success:function (result) {
22.            json = result;
23.            for (var i = 0; i < result.length; i++){
24.                triptypes.push(result[i].triptype);
25.                nums.push(result[i].nums);
26.                var ob = {name:"",value:""};
27.                ob.name = result[i].triptype;
28.                ob.value = result[i].nums;
29.                datatemp.push(ob);
30.            }
31.        },
32.        error :function(errorMsg) {
33.            alert("获取后台数据失败！");
34.        }
35.    });
36.    //指定图表的配置项和数据
37.    var option = {
38.        title: {
39.            text: '用户的旅游类型统计'
40.        },
41.        tooltip: {},
42.        legend: {
43.            data:['人次']
44.        },
45.        xAxis: {
46.            //结合
47.            axisLabel: {
48.                interval:0,
49.                rotate:20
50.            },
51.            data: triptypes
52.        },
53.        yAxis: {},
54.        series: [{
55.            name: '旅游类型',
56.            type: 'bar',
57.            //结合
58.            data: nums
59.        }]
60.    };
61.    //使用刚指定的配置项和数据显示图表
62.    myChart.setOption(option);
63.    //加载地图数据
64.    //旅游类型统计
```

```
65.        $.ajax({
66.            type:"GET",
67.            url:"/areastat",
68.            dataType:"json",
69.            async:false,
70.            success:function (result) {
71.                json = result;
72.                for (var i = 0; i < result.length; i++){
73.                    var ob = {name:"",value:""};
74.                    ob.name = result[i].area_name;
75.                    ob.value = result[i].nums;
76.                    datatemp.push(ob);
77.                }
78.            },
79.            error :function(errorMsg) {
80.                alert("获取后台数据失败!");
81.            }
82.        });
83.        var myChart2 = echarts.init(document.getElementById('main2'));
84.        option = {
85.            title: {
86.                text: '各地区酒店数量统计'
87.            },
88.            tooltip: {
89.                formatter:function(params,ticket, callback){
90.                    return params.seriesName+'<br />'+params.name+': '+params.value
91.                }
92.            },
93.            visualMap: {
94.                min: 0,
95.                max: 1500,
96.                left: 'left',
97.                top: 'bottom',
98.                text: ['高','低'],
99.                inRange: {
100.                    color: ['#e5e0e0', '#490104']
101.                },
102.                show:true
103.            },
104.            geo: {
105.                map: 'QD',
106.                roam: false,
107.                zoom:1.23,
108.                label: {
109.                    normal: {
110.                        show: true,
111.                        fontSize:'10',
112.                        color: 'rgba(0,0,0,0.7)'
113.                    }
```

```
114.                },
115.                itemStyle: {
116.                    normal:{
117.                        borderColor: 'rgba(0, 0, 0, 0.2)'
118.                    },
119.                    emphasis:{
120.                        areaColor: '#F3B329',
121.                        shadowOffsetX: 0,
123.                        shadowOffsetY: 0,
124.                        shadowBlur: 20,
125.                        borderWidth: 0,
126.                        shadowColor: 'rgba(0, 0, 0, 0.5)'
127.                    }
128.                }
129.            },
130.            series : [
131.                {
132.                    name: '酒店数量',
133.                    type: 'map',
134.                    geoIndex: 0,
135.                    data:datatemp
136.                }
137.            ]
138.        };
139.        $.getJSON('js/370200.json', function (geoJson) {
140.            myChart2.hideLoading();
141.            echarts.registerMap('QD', geoJson);
142.            myChart2.setOption(option);
143.        })
144.        //酒店星级分布情况统计
145.        var myChart3 = echarts.init(document.getElementById('main3'));
146.        var option3 = {
147.            title : {
148.                text: '酒店星级分布情况统计',
149.                subtext: '',
150.                x:'center'
151.            },
152.            tooltip : {
153.                trigger: 'item',
154.                formatter: "{a} <br/>{b} : {c} ({d}%)"
155.            },
156.            legend: {
157.                orient: 'vertical',
158.                left: 'left',
159.            },
160.            series : [
161.                {
162.                    name: '酒店类型',
163.                    type: 'pie',
164.                    radius : '55%',
```

```
165.            center: ['50%', '60%'],
166.            data:(function () {
167.                var datas = [];
168.                $.ajax({
169.                    type:"POST",
170.                    url:"/starstat",
171.                    dataType:"json",
172.                    async:false,
173.                    success:function (result) {
174.                        for (var i = 0; i < result.length; i++){
175.                            datas.push({
176.                                "value":result[i].nums, "name":result[i].stardetail
177.                            })
178.                        }
179.                    }
180.                })
181.                return datas;
182.            })(),
183.            itemStyle: {
184.                emphasis: {
185.                    shadowBlur: 10,
186.                    shadowOffsetX: 0,
187.                    shadowColor: 'rgba(0, 0, 0, 0.5)'
188.                }
189.            }
190.        }
191.        ]
192.    };
193.    myChart3.setOption(option3);
194. </script>
```

具体 Spring Boot 中 Controller、Service 和 DAO 层的代码不在本书讨论范围之内，读者可以参考配套的项目源码。其中基于酒店基本数据的统计分析部分包括：用户旅游类型分析、各地区酒店数量统计、酒店星级情况统计 3 个，可视化结果如图 14-5 所示。

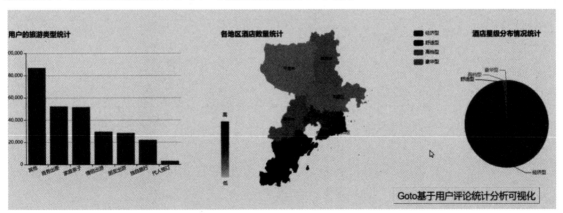

图 14-5

基于酒店用户评论数据的统计分析部分包括：网络人气酒店和用户满意度统计，可视化结果如图 14-6 所示。

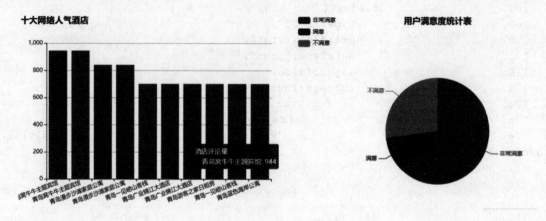

图 14-6